Collins · *do brilliantly*

Exam**Practice**

ASChemistry

Exam practice at its **best**

■ **George Facer**

■ **Series Editor: Jayne de Courcy**

I would like to dedicate this book to my father, John Facer, who taught me and inspired me with a lifelong love of Chemistry.

Published by Collins
An imprint of HarperCollins*Publishers*
77–85 Fulham Palace Road
Hammersmith
London
W6 8JB

Browse the complete Collins catalogue at
www.collinseducation.com

First published 2001
This revised edition published 2005

10 9 8 7 6 5 4 3 2 1

ISBN 0 00 719488 9

British Library Cataloguing in Publication Data
A Catalogue record for this publication is available from the British Library

Acknowledgements
Chris Conoley for reading and commenting on the manuscript
AQA examination questions are reproduced by permission of the Assessment and Qualification Alliance.
OCR examination questions are reproduced with the kind permission of OCR.
All solutions to questions taken from past examination papers are entirely the responsibility of the author. They have not been provided nor approved by the examination boards, and they may not necessarily constitute the only possible solutions.

Illustrations
Cartoon Artwork by Roger Penwill
DTP Artwork by Richard Morris
Artwork by Jerry Fowler

Every effort has been made to contact the holders of copyright material, but if any have been inadvertently overlooked, the Publishers will be pleased to make the necessary arrangements at the first opportunity.

Edited by Kathryn Senior and Rosamund Davies
Production by Katie Butler
Book Design by Bob Vickers and Gecko Limited
Printed and bound by Printing Express, Hong Kong

Contents

How this book will help you

by George Facer

Exam practice – how to answer questions better

This book will help you improve your performance in your AS Chemistry exam. It contains lots of **questions on the core topics of AS specifications.**

In exams, students often fail to gain the grades they are capable of, even when they have worked hard throughout their course and have a sound grasp of the facts. It is often not a lack of knowledge that leads to disappointing results, but poor examination technique. As an examiner, I often wish that I could sit alongside students as they answer questions, so that I could prompt them to make their explanation a bit fuller or ask them to clarify a point. **This book will help you with your examination technique so that you can make the most of your knowledge and score high marks in your exam.**

Each chapter in this book is broken down into four separate elements, aimed at giving you as much guidance and practice as possible:

1 Exam Question, Student's Answer and 'How to score full marks'

The questions and students' answers that I have chosen to start each chapter are typical ones. They show a number of mistakes frequently made by candidates under exam conditions.

The 'How to score full marks' section explains precisely where the student lost marks, e.g. not following instructions in the question, not putting in enough detail, etc. **I show you how to pick up those vital extra marks that make all the difference between an ordinary grade and a very good one.**

2 'Don't make these mistakes'

This section highlights the most common mistakes students make either in the exam itself or in their preparation for the exam. When you are into your last minute revision, you can quickly read through all of these sections and make doubly sure that you avoid these mistakes in your exam.

3 'Key points to remember'

These pages list **some of the most important facts you need to know or definitions you need to learn for each topic.** They are not meant to replace your notes, but are a quick check on those points that it is vital you know before going into your exam. You'll find these pages really helpful with last minute revision.

4 Questions to try, Answers and Examiner's comments

Each chapter ends with a number of exam questions for you to answer. Don't cheat. Sit down and answer the questions as if you were in an exam. Try to put into practice all that you have learnt from the previous sections in the chapter. I've included some exam hints before each question which should help you get the correct answers. Check your answers through and then look at the answers given at the back of the book. These are full mark answers.

In the 'Examiner's comments', I highlight anything tricky about the question which may have meant you did not get the correct answer. By reading through these sections, you can avoid making mistakes in your actual exam.

The topics covered by your specification

All exam boards have the same core of topics in AS Chemistry, but outside this core each has freedom about what they may put in their specification. In this book I have concentrated on including **questions on the areas that are common to all boards**. I have also, in some cases, added topics which are in the specification of two of them. Terminology varies from exam board to exam board and I have tried to take this into account as far as possible.

The chart below lets you see at a glance which chapters in this book are relevant to your particular specification.

	AQA	EDEXCEL	NUFFIELD	OCR	WJEC
Chapter 1 Atomic structure	Unit 1	Unit 1	Unit 1	Unit 2811	CH1
Chapter 2 Formulae, equations and amount of substance	Unit 1	Unit 1 and Unit 3B	Units 1 & 2	Unit 2811	CH1 and CH3a
Chapter 3 Structure and bonding	Unit 1	Unit 1	Ionic & metallic Unit 1 Covalent Unit 2	Unit 2811	CH1
Chapter 4 Periodic Table & Group II	Unit 1	Unit 1	Unit 1	Unit 2811	CH1
Chapter 5 Redox & Group VII	Unit 2	Unit 1	Unit 2	Unit 2811	CH1
Chapter 6 Energetics	Unit 2	Unit 2	Unit 1	Unit 2813/1	CH2
Chapter 7 Kinetics	Unit 2	Unit 2	Unit 2	Unit 2813/1	CH2
Chapter 8 Equilibrium	Unit 2	Unit 2	Unit 2	Unit 2813/1	CH2
Chapter 9 Organic chemistry	Unit 3	Unit 2	Alcohols Unit 1 Hydrocarbons & halogenoalkanes Unit 2	Unit 2812	CH2

(Note: AQA also includes the extraction of iron and aluminium in unit 2 and the distillation of petroleum and the cracking of a fraction in unit 3.
Edexcel includes industrial chemistry (the manufacture of ammonia, sulphuric acid, aluminium and chlorine) in unit 2.
OCR includes the cracking, isomerisation and reforming of fractions of crude oil in unit 2812.
These topics are not covered in this book due to lack of space.)

Exam Tips

Know the rubric. Do you have to answer all the questions? How many questions are there on the paper (this may vary from year to year)? How much time do you have for each question? As a rough guide the rate is about 1 mark per minute.

If a question is broken down into sections, **consider each section as part of the whole**, because the various parts will be linked.

Read each section of a question carefully and fully before answering it. **Highlight any command words** (see below).

The total marks for each part will be shown in or near the margin. **If there are two marks for a part, make sure that you write at least two statements that are worthy of scoring one point each**.

Set out calculations clearly, showing what you are calculating at each step. Give your final answer to the correct number of significant figures
(if in doubt give it to 3) and add the unit. Don't round up intermediate answers to 1 or 2 significant figures.

If you want to alter something, don't use Tippex. Neatly cross it out and write your answer in a free space in the question paper. If it is on a different page, alert the examiner by writing 'see page xx'. If you change your mind and want the examiner to mark what you originally wrote, write 'please ignore crossing out'.

If you are asked for a reagent or name, **don't give alternatives**. If one is wrong, you will be penalised.

When you have finished, **check as many of your answers as you have time for**:

- **Start with a calculation**. Have you worked out the relative molecular mass correctly? Have you misread any numbers? Re-calculate the answer to check that you have not made a calculator error. Finally check the number of significant figures and the units of your answer.
- **Check all chemical formulae**. Make sure that you have not made any silly errors such as $KMnO_4^-$ (this is surprisingly common). Check the oxidation state in names.
- **Check that all equations balance**.
- **Beware of contradictory statements** such as conc H_2SO_4(aq) or NaOH in acid.
- Check that you have given **the full name or formula** for a reagent.
- In organic chemistry **check formulae** to ensure that all carbon atoms have 4 bonds, all oxygen atoms 2 and hydrogen only 1.

Make sure you really understand the meaning of these 'command words':

- **Define** – This means that you must recall the formal textbook definition. It is often helpful to include an example or equation to supplement your definition.
- **Explain** – This requires a detailed answer. Look to see how many marks there are, and make at least the same number of points.
- **State** – A simple, often one-word answer is all that is required. There is no need for any explanation.
- **Deduce** – Use the data supplied to answer the question.
- **Hence deduce** – Use the answer that you have just obtained in the previous part to work out the answer to this part.
- **Suggest** – You are not expected to have learnt the answer. You should be able to apply your knowledge of similar reactions or substances and work out the answer.
- **Compare** – You must say something about *both* substances or reactions.
- **Name** – Give the full name not the formula.
- **Structural formula** – It is safest to give a formula showing all the atoms and all the bonds. Never use added up formula when it is ambiguous, e.g. for the C_3H_7 group.

Exam Question and Student's Answer

1 This question is about three elements, **A**, **B** and **C** in the Periodic Table.

(a) Each atom of **A** has 15 electrons.

(i) Identify element **A**.

Element A is phosphorus ✓

[1 mark] 1/1

(ii) Complete the electronic configuration of an atom of element **A**.

$1s^2$ $2s^2$ $2p^6$ $3s^2$ $3p^3$ ✓

[1 mark] 1/1

(iii) Predict the charge on an ion of **A** and complete its electronic configuration.

Charge on ion 3^+ ✗

Electronic configuration of an ion of **A** $1s^2$ $2s^2$ $2p^6$ $3s^2$ $3p^6$ ✓

[2 marks] 1/2

(b) Element **B** exists as a mixture of three isotopes.

(i) What is the difference between the atomic structures of isotopes?

They have a different number of neutrons in their nucleus. ✓

[1 mark] 1/1

(ii) The atoms of element **B** have eight electrons in their 3d sub shell.

Identify element **B**.

It is argon. ✗

[1 mark] 0/1

(c) A sample of element **B** was analysed in a mass spectrometer. The relative atomic mass of element **B** can be calculated from the results shown in the table below.

	isotope 1	isotope 2	isotope 3
relative isotopic mass	58.0	60.0	62.0
percentage composition	68.2	27.3	4.5

(i) Explain what is meant by the term relative atomic mass.

It is the average mass of the element compared with 1/12th the mass of a carbon 12 atom. ✓

[3 marks] 1/3

(ii) Explain what is meant by the term relative isotopic mass.

It is the average ✗ mass of the atoms of that isotope ✓ relative to 1/12th the mass of a carbon 12 isotope.

[2 marks] 1/2

(iii) Using the information in the table above, calculate the relative atomic mass of this sample of **B**, to the correct number of significant figures.

$$A_r = \frac{(58.0 \times 68.2) + (60.0 \times 27.3) + (62.0 \times 4.5)}{100} = 58.726$$ ✓ ✗

[2 marks] 1/2

(d) The table below shows the first five successive ionisation energy values for the element **C**.

Ionisation energy	Value / kJ mol^{-1}
1st	577
2nd	1820
3rd	2740
4th	11600
5th	14800

(i) Define the term 2nd ionisation energy.

It is the energy required to remove two ✗ electrons from a gaseous ✓ atom of the element.

[3 marks] 1/3

(ii) Use the data in the table above and the Periodic Table, to suggest the identity of element **C**.

Element C is probably aluminium. ✓

[1 mark] 1/1

9/17

[Total 17 marks]

How to score full marks

(a) (iii) Atoms in the s and p blocks gain or lose electrons so as to reach the electronic configuration of a noble gas. The student correctly realised that the atom had to gain 3 electrons, but thought that this made the ion positive. **Electrons are negative and so gaining electrons makes the atom negative.**

(b) (i) As the question asks about the atomic structures, you must state that **the number of neutrons are different** rather than that their masses are different. There is no need to state that they have the same number of protons.

(ii) The student misread the question and gave the answer as if **B** had eight electrons in its 3rd orbit, not eight in the 3d sub-orbit. The element is **nickel**.

(c) (i) You must make it clear that the average takes into account the relative amounts of each isotope. The phrase 'weighted average' does this. Also it is the mass of the **atoms** not of the element that is compared with $1/12^{th}$ the mass of an atom of carbon 12.

(ii) There is **no averaging** when defining relative isotopic mass. It is simply the mass of 1 atom of that isotope relative to the mass of 1 atom of carbon 12.

(iii) You must give your answer **to the correct number of significant figures**. Here all the data is given to 3 sf and so your answer must also be given to 3 sf. Decimal places are not relevant (except in pH calculations).

(d) (i) The answer is wrong. The 2nd ionisation energy is the **energy required per mole to remove one electron from a 1+ gaseous ion**. This is for the process

$$C^+(g) \rightarrow C^{2+}(g) + e^-$$

(ii) The student correctly identified the **big jump between the 3rd and 4th ionisation energies**. This is when an electron has to be removed from an inner shell, and so element **C** has 3 electrons in its outer shell and so is in Group 3.

Don't make these mistakes ...

In your definition of an s-block element, do **not** state that 'an s-block element has a **partially** filled s orbital'. Beryllium is an s-block element but its s orbital is full; its electronic configuration is $1s^22s^2$. The reason that beryllium is an s-block element is that its **highest energy** electron is in an s orbital. (The same rule holds true for deciding whether an element is a *p*- or *d*-block element.)

Do not forget to give an **equation, including state symbols**, in your definition of ionisation energy and electron affinity. This may help you to gain full marks if you leave out a scoring point such as the fact that the atoms must be gaseous.

Mass spectra are caused by **positive** ions. Do **not** leave out the '+ sign' in the formula of an ion that causes a specific line in a mass spectrum.

Ionisation energies are always related to the **removal** of an electron. For chlorine the process

$$Cl(g) + e^- \rightarrow Cl^-(g)$$

is called the (first) **electron affinity** of chlorine.

In your explanation of isotopes do **not** imply that they are different **elements**, but stress that they are different **atoms** of the **same** element.

Key points to remember

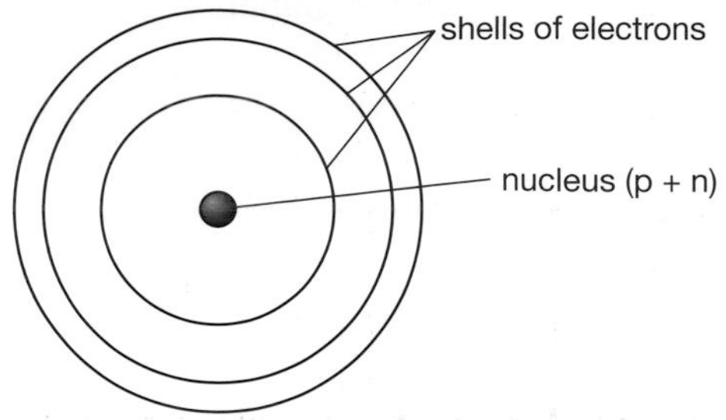

Atomic (or proton) number (Z) is the number of protons in the nucleus.

Mass number (A) of an isotope is the sum of the number of protons and neutrons in the nucleus.

Orbit (shell)	orbitals	max no of electrons
1	s	2
2	s & p	2 + 6 = 8
3	s, p & d	2 + 6 + 10 = 18

The order of filling orbitals is

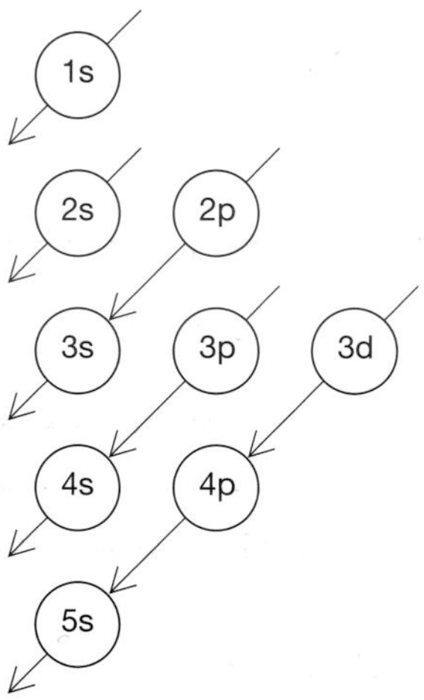

Electron structures are:

$_{8}O$	$1s^2,2s^2\ 2p^4$	
$_{10}Ne$	$1s^2,2s^2\ 2p^6$	
$_{17}Cl$	$1s^2,2s^2\ 2p^6,3s^2\ 3p^5$	or $[Ne]\ 3s^2,3p^5$
$_{18}Ar$	$1s^2,2s^2\ 2p^6,3s^2\ 3p^6$	
$_{20}Ca$	$1s^2,2s^2\ 2p^6,3s^2\ 3p^6,\ 4s^2$	or $[Ar]\ 4s^2$
$_{26}Fe$	$1s^2,2s^2\ 2p^6,3s^2\ 3p^6\ 3d^6,\ 4s^2$	or $[Ar]\ 3d^6,\ 4s^2$

Mg is *s*-block because its electronic structure is $1s^2\ 2s^2\ 2p^6\ 3s^2$

Ar is *p*-block because its electronic structure is $1s^2\ 2s^2\ 2p^6\ 3s^2\ 3p^6$

Ti is *d*-block because its electronic structure is $1s^2\ 2s^2\ 2p^6\ 3s^2\ 3p^6\ 3d^2\ 4s^2$ (3d is of higher energy than 4s).

An ***s*-block element** is one where the highest occupied energy level is an s orbital.

A ***p*-block element** is one where the highest occupied energy level is a p orbital.

A ***d*-block element** is one where the highest occupied energy level is a d orbital.

Carbon has three naturally occurring isotopes: ^{12}C, ^{13}C and ^{14}C

Mass number / Atomic number	$^{12}_{6}C$	$^{13}_{6}C$	$^{14}_{6}C$
No of protons	6	6	6
No of neutrons	6	7	8

Isotopes are different atoms of the same element. The have the same number of protons, but a different number of neutrons. Thus they have different masses.

The **relative atomic mass** of an element is the average of the masses of the naturally occurring isotopes (taking into account their abundances) divided by $\frac{1}{12}$th the mass of a carbon 12 atom.

Chlorine has two isotopes:

^{35}Cl with an abundance of 75%, and

^{37}Cl with an abundance of 25%.

The average mass is $\frac{(35 \times 75 + 37 \times 25)}{100} = 35.5$

Mass spectra are caused by a gaseous substance being bombarded by high energy particles (electrons). The substance forms positive ions that are accelerated by an electric field and then deflected by a magnetic field. The stream of ions is split up according to the $\frac{m}{z}$ (mass/charge) value of the ions. The ions produced may be molecular ions or fragments (so bromine will have two lines at $\frac{m}{z}$ of 79 and 81 as well as those shown in the figure opposite).

The **mass spectrum of molecular bromine is**:

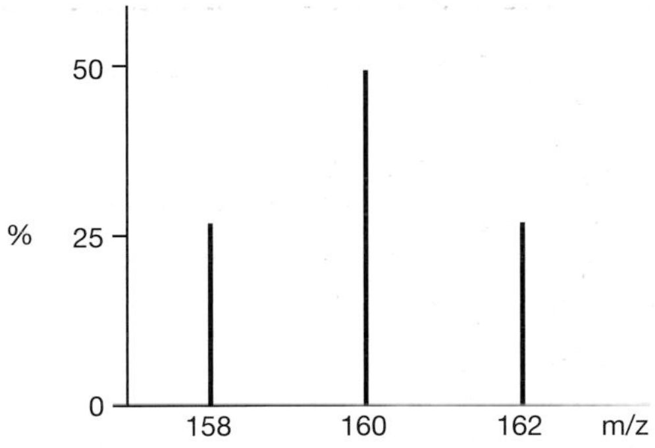

$\frac{m}{z}$ = 158 is caused by $(^{79}Br\ ^{79}Br)^+$

$\frac{m}{z}$ = 160 is caused by $(^{79}Br\ ^{81}Br)^+$

$\frac{m}{z}$ = 162 is caused by $(^{81}Br\ ^{81}Br)^+$

As the lines at $\frac{m}{z}$ of 158 and 162 are of equal height, there is 50% of each isotope.

The **1st ionisation energy** is the energy required to remove one electron from one mole of gaseous atoms.

The 2nd ionisation energy is the energy required to remove one electron from one mole of gaseous ions with a single positive charge.

The value of the 1st ionisation energy depends mainly upon the pull of the nucleus on the electron that is removed (the stronger the pull, the larger the ionisation energy). This pull is determined by two factors:

- The effective nuclear charge (this equals the atomic number minus the number of inner or shielding electrons);
- The distance from the nucleus to the electron (the smaller the atom, the larger the ionisation energy).

The **1st ionisation energy** of chlorine is the energy per mole for

$$Cl(g) \rightarrow Cl^+(g) + e^-$$

The **2nd ionisation energy** of chlorine is the energy per mole for

$$Cl^+(g) \rightarrow Cl^{2+}(g) + e^-$$

Ionisation energies are always positive (endothermic).

Across a period the **nuclear charge increases** but there is no increase in shielding since the number of inner electrons remains the same. Also, the atoms get slightly smaller across a period. These factors cause a general increase in the ionisation energy across a period.

Down a group, the nuclear charge increases, but so does the number of shielding electrons. Thus, the effective nuclear charge remains the same, but the outer electron becomes further from the nucleus and so is easier to remove.

The 1st electron affinity for chlorine is the energy per mole for:

$$Cl(g) + e^- \rightarrow Cl^-(g)$$

1st electron affinities are always **negative** (exothermic).

The 1st electron affinity is the energy change when 1 electron is added to each of one mole of gaseous atoms.

Questions to try

Examiner's hints ***for question 1***

(a) Elements are classified into blocks according to the highest occupied energy level.
Ionisation energies refer to the **loss** of electrons. Their values depend on the strength of the force of attraction between the nucleus and the electron being removed. This is determined by:

- the charge on the nucleus;
- to what extent the nucleus is shielded by inner electrons;
- the distance between the nucleus and the electron being removed;
- any repulsion between two electrons in a given orbital.

Q1

Antimony (Sb) and nitrogen (N) are both group 5 elements.

(a) (i) Write the electronic structures, using the s, p, d notation, of nitrogen.

....

[1 mark]

(ii) Why is nitrogen classified as a *p*-block element?

....

[1 mark]

(iii) Write the equation for the first ionisation energy of nitrogen.

....

....

....

[2 marks]

(iv) Why is the first ionisation energy of nitrogen more than that of oxygen but less than that of fluorine?

....

....

....

....

....

[4 marks]

(b) Antimony has isotopes of mass numbers 121 and 123, and a relative atomic mass of 121.8.

(i) Distinguish between the terms **mass number** and **relative atomic mass.**

Mass number ....

....

....

Relative atomic mass ....

....

....

[5 marks]

(ii) The mass spectrum of antimony is shown below. Use the data to show why the relative atomic mass of antimony is 121.8.

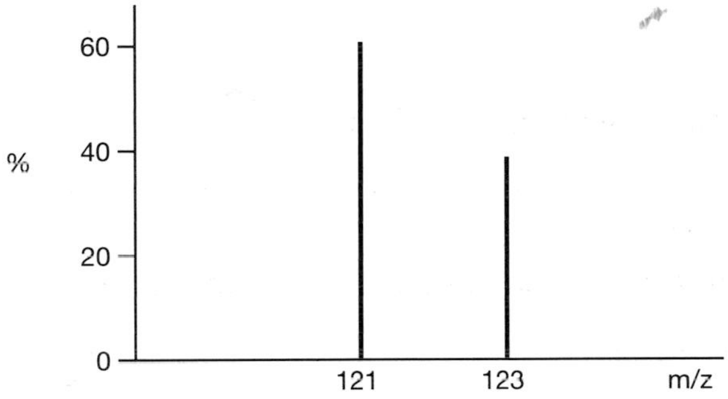

[2 marks]

(iii) In a mass spectrometer the atoms of an element are bombarded by high energy electrons and form positive ions. Describe how these ions are subsequently separated according to their mass/charge $\left(\frac{m}{z}\right)$ ratio.

....

....

....

[3 marks]

[Total 18 marks]

Examiner's hints ***for question 2***

(a) Answer the three parts of **(a)** separately:

- List the relative masses and charges.
- Discuss the numbers of each particle in the different isotopes and the resulting difference in properties.
- Think what part of an atom determines chemical properties.

(b) Start with the element being placed in the mass spectrometer. What is done to it initially and what then happens to the particles? Finally what causes the isotopes to be separated?

Q2

(a) State the relative charge and relative mass of a proton, of a neutron and of an electron. In terms of particles, explain the relationship between two isotopes of the same element. Explain why these isotopes have identical chemical properties.

....

....

....

....

....

....

....

.... [7 marks]

(b) A mass spectrometer can be used to find the relative abundance and the relative isotopic masses of the isotopes in an element. Describe the principles of a mass spectrometer and how the isotopes are separated in the spectrometer.

....

....

....

....

....

....

....

....

....

.... [8 marks]

[Total 15 marks]

The answers to these questions are on pages 77–78.

[Relative atomic masses required for questions in this chapter are:
H=1, C = 12, N = 14, O = 16, Na = 23, S = 32, Cl = 35.5, Ca = 40 and Ag = 108]

Exam Questions and Student's Answers

1 (a) A 3.30 g sample of silver nitrate, $AgNO_3$, was dissolved in water to produce 50.0 cm^3 of silver nitrate solution.

Calculate the concentration of this solution in mol dm^{-3}.

The molar mass of $AgNO_3$ = 108 + 14 + (3 x 16) = 170 g mol^{-1} ✓
Amount of $AgNO_3$ = mass / molar mass = 3.30g/170 g mol^{-1} = 0.0194 mol ✓
Concentration = 0.0194 mol ÷ (50.0 / 1000) dm^3 = 0.388 mol dm^{-3} ✓

[3 marks] 3/3

(b) A sample of impure silver nitrate was heated and the following reaction occurred:

$$2AgNO_3(s) \rightarrow 2Ag(s) + 2NO_2(g) + O_2(g)$$

(i) If 32.5 cm^3 of oxygen gas was given off, calculate the mass of silver nitrate in the impure sample.
(One mole of gas occupies a volume of 24.0 dm^3 under the conditions of the experiment.)

moles of oxygen = volume / molar volume = 32.5 / 24.0 = 1.35 mol ✗
moles of silver nitrate = 1.35 mol ✗
mass of silver nitrate = moles x molar mass = 1.35 mol x 170 g mol^{-1}
= 230 g ✓ (ecf)

[3 marks] 1/3

(ii) Calculate the number of oxygen atoms in the 32.5 cm^3 of oxygen gas. (The Avogadro constant = 6.02×10^{23} per mol.)

Number of oxygen atoms = moles x Avogadro's constant
= 1.35 x 6.02 x 10^{23}
= 8.13 x 10^{23} atoms ✓ (ecf)

[2 marks] 1/2

(c) Silver nitrate solution reacts with calcium chloride solution to give a precipitate of silver chloride.
Write the ionic equation, with state symbols, for this reaction.

$2AgNO_3(aq) + CaCl_2(aq) \rightarrow 2AgCl(s) + Ca(NO_3)_2(aq)$
$2Ag^+(aq) + 2NO_3^-(aq) + Ca^{2+}(aq) + 2Cl^-(aq) \rightarrow 2AgCl(s) + 2NO_3^-(aq) + 2Ca^{2+}(aq)$
Equation ✗ state symbols ✓

[2 marks] 1/2

(d) Another silver compound was analysed and found to contain the following percentages by mass.

Element	% by mass
Silver	78.3
Carbon	4.3
Oxygen	17.4

Calculate the empirical formula of this silver compound.

Element	%	% ÷ A_r	÷ smallest ✓	X 2
Silver	78.3	÷ 108 = 0.725	1.0	2
Carbon	4.3	÷ 6 ✗ = 0.717	1.0	2
Oxygen	17.4	÷ 16 = 1.09	1.5	3

The empirical formula is $Ag_2C_2O_3$ ✓ (ecf)

[3 marks] 2/3

[Total 13 marks] 8/13

2 (a) Calculate the number of moles of:

(i) hydrogen in 68 cm^3 of H_2 gas at room temperature and pressure.
(1 mol of gas occupies 24 dm^3 at room temperature and pressure.)

moles $H_2 = \frac{68}{24}$ ✗ $= 2.8$ mol

[1 mark] 0/1

(ii) sodium hydroxide in 33 g NaOH.

moles NaOH $= \frac{33}{40} = 0.83$ mol ✓

[1 mark] 1/1

(iii) sodium hydroxide in 27.3 cm^3 of 0.123 mol dm^{-3} sodium hydroxide solution.

moles NaOH $= \frac{27.3}{1000} \times 0.123 = 3.36 \times 10^{-3}$ ✓

[1 mark] 1/1

(b) Calculate the mass of $Na_2CO_3.10H_2O$ crystals needed to make 250 cm^3 of a 0.164 mol dm^{-3} solution of sodium carbonate.

250 cm^3 is $\frac{1}{4}$ of 1 dm^3

moles required $= \frac{1}{4} \times 0.164 = 0.0410$ ✓

mass = moles X M_r ✓ = 0.0410 X 106 ✗ = 4.35 g ✗

[3 marks] 2/3

(c) In a titration 25.0 cm^3 of the 0.164 mol dm^{-3} solution of sodium carbonate required 28.7 cm^3 of a solution of hydrochloric acid. They react according to the equation:

$$Na_2CO_3(aq) + 2HCl(aq) \rightarrow 2NaCl(aq) + CO_2(g) + H_2O(l)$$

Calculate the concentration of the hydrochloric acid solution:

(i) in mol dm^{-3}
(ii) in g dm^{-3}

(i) volume of $Na_2CO_3(aq)$ solution $= \frac{25.0}{1000} = 0.0250\ dm^3$

moles of Na_2CO_3 $= 0.0250 \times 0.164 = 4.10 \times 10^{-3}$ ✓

moles of HCl $= 4.10 \times 10^{-3}$ ✗

concentration $= 4.10 \times 10^{-3} \times 0.0287 = 1.18 \times 10^{-4}$ ✗ 1/3

(ii) molar mass of HCl = 1 + 35.5 = 36.5 g mol^{-1}

concentration in g dm^{-3} $= 36.5 \times 1.18 \times 10^{-4}$

$= 4.29 \times 10^{-3}$ g dm^{-3} ✓ [4 marks] 1/1

[Total 10 marks] 6/10

How to score full marks

1 (a) The student set out the calculation clearly, with **the units** shown during the calculation.

(b) (i) A common error was made here. The volume of oxygen was given in cm^3 and, as the molar volume was given in dm^3, **a conversion must be made**, either of 24.0 dm^3 to 24000 cm^3 or 32.5 cm^3 to 0.0325 dm^3. The second error was to fail to use the stoichiometry of the reaction. The ratio of O_2 to $AgNO_3O_2$ is 2:1.

Examiners mark by processes. In this case the last step was correctly performed based upon the wrong value of moles of silver nitrate. This is why **ecf (error carried forward)** was put next to the ✓. The correct answer is:

Amount of oxygen gas = volume / molar volume = 0.0325 dm^3 / 24.0 dm^3 mol^{-1}
= 0.00135 mol ✓

Amount of silver nitrate = **2** × 0.00135 = 0.00270 mol ✓

Mass of silver nitrate in sample = 0.00270 mol × 170 g mol^{-1} = 0.459 g ✓ (allow 0.460 g)

(ii) The question asked for the number of **atoms** not molecules of oxygen, so the correct process is 2 × moles. Again the examiner gave an ecf mark for multiplying the moles by the Avogadro constant, 6.02×10^{23} per mole.

(c) **The student lost a mark for failing to cancel the spectator ions** that, in this equation, are $Ca^{2+}(aq)$ and $2NO_3^-(aq)$.

(d) The student made a common error of carelessly using the atomic number of carbon (6) and not its **relative atomic mass** (12). If 12 had been used, the answer would have been Ag_2CO_3.

2 (a) (i) The correct answer is $\frac{68}{1000} \div 24 = 0.0028$ mol.

The molar volume (24 dm^3 at room temperature and pressure) is given in dm^3, **so you must first convert the volume of 68 cm^3 to dm^3 by dividing by 1000.**

(b) The last line of the answer contains a mistake. The relative molecular mass of the crystals, which have the formula $Na_2CO_3.10H_2O$, is $[2 \times 23 + 12 + 3 \times 16 + 10 \times (2 + 16)] = 286$, so the mass needed is $0.0410 \times 286 = 11.7$ g.

(c) (i) The first mark is awarded for calculating the moles of sodium carbonate correctly using the **concentration and volume.**

The second line is wrong. The equation has a ratio of 2HCl for each Na_2CO_3, so moles of HCl = $\frac{2}{1} \times$ moles of $Na_2CO_3 = \frac{2}{1} \times 4.10 \times 10^{-3} = 8.20 \times 10^{-3}$

The third line is also wrong. **Concentration is moles/volume** and so the 8.20×10^{-3} must be **divided** by 0.0287.

$\therefore$ concentration of HCl solution = $\frac{8.20 \times 10^{-3}}{0.0287} = 0.286$ mol dm^{-3}.

(ii) **The mark was awarded here for the correct chemical process.** To go from concentration in moles dm^{-3} to g dm^{-3}, you must multiply by the relative molecular mass, and the student here did this correctly.

Don't make these mistakes ...

Don't forget to check that your **equation** is correct (see the Key Points section on pages 19–20).

Don't round up numbers in the middle of a calculation and don't give your final answer to the wrong number of significant figures.

The **molar volume** of a gas is sometimes given in dm^3 mol^{-1} and sometimes in cm^3 mol^{-1}. Your gas volume must be in the same units.

Don't forget the **stoichiometry** (the numbers of each substance in the equation). To convert moles of chemical A to moles of chemical B, the following has to be calculated:

moles of B = moles of A x the ratio B/A in the equation.

When a **concentration** is in mol dm^{-3} (sometimes called molarity, symbol M), **any volume must be in dm^3**, so convert a volume given in cm^3 to dm^3 by dividing by 1000.

Don't confuse **volume units.**

Don't write down a series of numbers in a calculation. **Set your work out** showing what chemical each number refers to, for example:

amount of $Fe_2O_3 = \frac{\text{mass}}{M_r} = \frac{1.75}{160}$

= 0.01094 mol

or % Fe = 37.1%

Amount of substance

$$\text{Moles (n)} = \frac{\text{mass (m)}}{\text{relative molecular mass}}$$

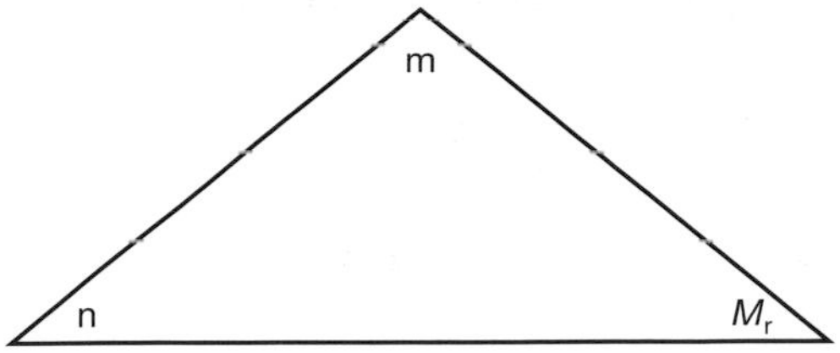

What is the number of moles of Na in 1.6 g of Na?

$$\frac{\text{mass}}{M_r} = \frac{1.6}{23} = 0.070$$

How many moles of Na in 1.6g of NaOH?

$$\frac{\text{mass}}{M_r} = \frac{1.6}{40} = 0.040$$

$$\text{Moles (n) of a gas} = \frac{\text{volume of gas}}{\text{molar volume}}$$

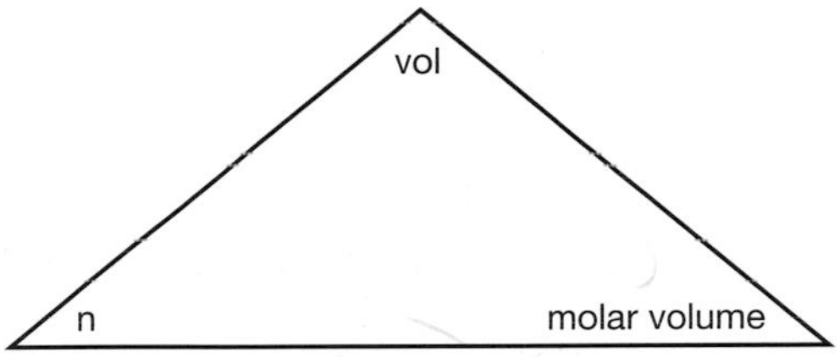

The molar volume of a gas is 22.4 dm^3 at standard temperature and pressure or 24 dm^3 at room temperature and pressure.

How many moles of $O_2(g)$ in 1.6 dm^3 at STP?

$$\text{volume/molar volume} = \frac{1.6}{22.4} = 0.071$$

How many moles of $O_2(g)$ in 1.6 dm^3 at RTP?

$$\text{volume/molar volume} = \frac{1.6}{24} = 0.067$$

Moles (n) of solute in a solution =

$$\text{concentration in mol dm}^{-3} \times \frac{\text{volume in cm}^3}{1000}$$

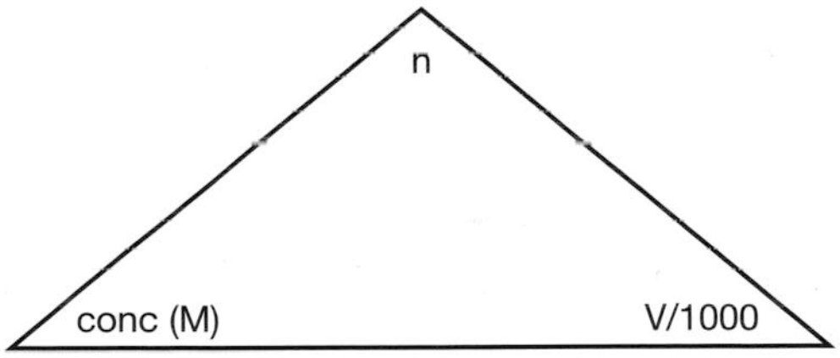

How many moles of NaOH in 23.4 cm^3 of 0.123 mol dm^{-3} solution?

$$\text{concentration} \times \text{V(in dm}^3) = 0.123 \times \frac{23.4}{1000}$$

$= 2.88 \times 10^{-3}$

What is the concentration of the solution produced when 0.22 mol of NaOH is dissolved in 0.11 dm^3 of water?

$$\text{moles/volume(in dm}^3) = \frac{0.22}{0.11} = 2.0 \text{ mol dm}^{-3}$$

Formulae and equations

Empirical formulae

These are the simplest whole number ratio of atoms in the compound. They can be calculated from % composition.

Give values in column 2 to 2 decimal places.

Give values in column 3 to 1 decimal place.

If values in column 3 end in 0.5 multiply all by 2.

Calculate in three columns:

Element and %	% ÷ A_r	÷ smallest
C 48.7%	÷ 12 = 4.06	÷ 2.7 = 1.5
H 8.1%	÷ 1 = 8.10	÷ 2.7 = 3.0
O 43.2%	÷ 16 = 2.70	÷ 2.7 = 1.0

Multiply all by 2 because one ends in 0.5

The empirical formula is $C_3H_6O_2$

Key points to remember

Calculations

There are three main types of calculation:

mass of A → mass of B
mass of A → volume of gas B (or vice versa)
titration of A with B

For all three follow the same route

1 Write a balanced equation.
2 Calculate the moles of A.
3 Moles of B = moles of A × the ratio $\frac{B}{A}$.
4 Calculate the answer.

For steps 2 and 4 use the formula shown above under the heading amount of substance.

For step 3 the ratio $\frac{B}{A}$ is the ratio of the numbers of molecules of B/number of molecules of A in the balanced equation that you wrote in step 1.

Now let's see some examples:

(a) Calculate the mass of H_2SO_4 that will react with 1.25 g of Al.

1 $2Al + 3H_2SO_4 \rightarrow Al_2(SO_4)_3 + 3H_2$

2 moles of Al = $^{1.25}/_{27}$ = 0.0463

3 moles of $H_2SO_4 = 0.0463 \times {}^3/_2 = 0.0694$

4 mass of $H_2SO_4 = 98 \times 0.0694 = 6.81$ g

(b) Calculate the volume of CO_2(at STP) that will react with 5.21 g of NaOH.

1 $2NaOH + CO_2 \rightarrow Na_2CO_3 + H_2O$

2 moles of NaOH = $\frac{5.21}{40} = 0.130$

3 moles of $CO_2 = 0.130 \times \frac{1}{2} = 0.0651$

4 volume of $CO_2 = 22.4 \times 0.0651 = 1.46\ dm^3$

(c) Calculate the volume of 0.333 mol dm^{-3} NaOH that will react with 22.2 cm^3 of 0.0555 mol dm^{-3} H_2SO_4 solution

1 $2NaOH + H_2SO_4 \rightarrow Na_2SO_4 + 2H_2O$

2 moles of $H_2SO_4 = 0.0555 \times \frac{22.2}{1000} = 1.23 \times 10^{-3}$

3 moles of NaOH = $1.23 \times 10^{-3} \times \frac{2}{1}$

4 volume of NaOH = $\frac{2.46 \times 10^{-3}}{0.333} = 7.39 \times 10^{-3} dm^3$

$= 7.39\ cm^3$

Molecular equations

Check that all the formulae are correct.

Check that there are the same number of atoms of each element on each side.

If not change the numbers in front of the formulae until the equation balances.

Do not alter any formulae.

Consider this reaction as an example:

Calcium nitrate decomposes into calcium oxide, nitrogen(IV) oxide and oxygen.

Try: $Ca(NO_3)_2 \rightarrow CaO + 2NO_2 + O_2$

Here is the check list:

Formulae: all correct

Numbers of atoms on each side:

Ca 1 and 1, N 2 and 2, O 6 and 7.

The equation does not balance.

Now try $2Ca(NO_3)_2 \rightarrow 2CaO + 4NO_2 + O_2$

Check again:

Ca 2 and 2, N 4 and 4, O 12 and 12

The equation now balances.

Molecular formulae

These are the exact numbers of each atom in the molecule.

If an empirical formula is CH_2 and the relative molecular mass is 56, the molecular formula is C_4H_8, because CH_2 has a mass of 14 and $14 \times 4 = 56$.

Ionic equations

Apart from following the rules for molecular equations, you must also check that the total charge on the left equals the total charge on the right.

Chlorine oxidises Fe^{2+} ions to Fe^{3+} ions:

Try $Cl_2 + Fe^{2+} \rightarrow 2Cl^- + Fe^{3+}$

The formulae are correct, numbers of atoms balance, but the charge on left is 2+ while the charge on right is 2– + 3+ = 1+.

The correct equation is:

$Cl_2 + 2Fe^{2+} \rightarrow 2Cl^- + 2Fe^{3+}$

Now charge: left = 4+, right is 2– + 6+ = 4+

Questions to try

Examiner's hint ***for question 1***

(a) moles = $\frac{\text{mass}}{M_r}$ or for a gas, moles = $\frac{\text{volume}}{\text{molar volume}}$

moles of solute in a solution = $\frac{\text{volume (V) in cm}^3}{1000}$ × concentration (M or molarity) in mol dm^{-3}

Q1

Calcium and hydrochloric acid react according to the equation:

$$Ca(s) + 2HCl(aq) \rightarrow CaCl_2(aq) + H_2(g)$$

(a) 0.122 g of calcium metal was added to 25.0 cm^3 of 1.23 mol dm^{-3} solution of hydrochloric acid and the hydrogen gas was collected at room temperature and pressure.

(i) Calculate whether the calcium or the hydrochloric acid was in excess.

....

....

....

.... [4 marks]

(ii) Hence calculate the volume of hydrogen gas produced. (1 mol of gas under these conditions occupies 24 000 cm^3.)

....

....

[2 marks]

(b) Calcium metal often contains an impurity of calcium oxide on its surface. This also reacts with hydrochloric acid, but does not produce a gas. In the experiment in (a), it was found that only 70.1 cm^3 of hydrogen was produced. Calculate the percentage purity of the calcium metal.

....

....

[2 marks]

(c) Hard water contains calcium ions. The amount can be found by adding an excess of sodium ethandioate and weighing the precipitate formed. Sodium ethandioate contains the following percentages by mass:

sodium 34.3% carbon 17.9% oxygen 47.8%

(i) Calculate the empirical formula of sodium ethandioate.

....

....

....

[3 marks]

(ii) The relative molecular mass of sodium ethandioate is 134. Use this to deduce its molecular formula.

....

....

[2 marks]

[Total 13 marks]

Examiner's hints ***for question 2***

(b) (ii) Don't forget the stoichiometry of the reaction. Be careful about working out the molar mass of ammonium sulphate.

(c) You will need to double the first equation so that the $NaHCO_3$ cancels on addition of the two equations. This will give you the ratio of NaCl to Na_2CO_3.

(d) (i) Balance the hydrogen last.

Q2

(a) A solution containing 0.732 mol of ammonia was made up to 250 cm^3 in a volumetric flask by adding water. Calculate the concentration of ammonia in this final solution and state the appropriate units.

....

.... [2 marks]

(b) A different solution of ammonia was reacted with sulphuric acid as shown in the equation below.

$$2NH_3(aq) + H_2SO_4(aq) \rightarrow (NH_4)_2SO_4(aq)$$

In a titration, 25.0 cm^3 of a 1.24 mol dm^{-3} solution of sulphuric acid required 30.8 cm^3 of this ammonia solution for complete reaction.

(i) Calculate the concentration of ammonia in this solution.

....

....

(ii) Calculate the mass of ammonium sulphate in the solution at the end of this titration.

....

....

....

[6 marks]

(c) Sodium carbonate is manufactured from ammonia in a two-stage process as shown by the equations below.

$$NaCl + NH_3 + CO_2 + H_2O \rightarrow NaHCO_3 + NH_4Cl$$

$$2NaHCO_3 \rightarrow Na_2CO_3 + H_2O + CO_2$$

Calculate the maximum mass of sodium carbonate which could be obtained from 800 g of sodium chloride.

....

....

[4 marks]

(d) Ammonia and magnesium hydroxide are formed when magnesium nitride reacts with water.

(i) Balance the equation, given below, for the reaction between magnesium nitride and water.

$$Mg_3N_2 + \quad H_2O \rightarrow \quad Mg(OH)_2 + \quad NH_3$$

(ii) Calculate the number of moles, and hence the number of molecules, of NH_3 in 0.263 g of ammonia gas.
(The Avogadro constant, L, = 6.02×10^{23} mol^{-1})

....

....

....

[3 marks]

[Total 15 marks]

Examiner's hint ***for question 3***

When balancing an equation:

1. Check there are the same numbers of each element on each side of the equation.
2. For ionic equations, check that the total charge on the left-hand side of the equation equals the total charge on the right-hand side.

If either is wrong, alter the numbers in front of the formulae until the equation balances.

Do not alter formulae.

Q3

Balance the following equations

(a) $Al_2O_3 + \quad HCl \rightarrow \quad AlCl_3 + \quad H_2O$ [1 mark]

(b) $LiNO_3 \rightarrow \quad Li_2O + \quad NO_2 + O_2$ [1 mark]

(c) $K + \quad H_2O \rightarrow \quad KOH + \quad H_2$ [1 mark]

(d) $Ca(OH)_2 + Na_3PO_4 \rightarrow Ca_3(PO_4)_2 + NaOH$ [2 marks]

(e) $Fe^{3+}(aq) + Sn^{2+}(aq) \rightarrow Fe^{2+}(aq) + Sn^{4+}(aq)$ [1 mark]

(f) $I^-(aq) + IO_3^-(aq) + H^+(aq) \rightarrow I_2(aq) + H_2O(l)$

[2 marks]

[Total 8 marks]

The answers to these questions are on pages 78–80.

Exam Questions and Student's Answers

1 (a) An ammonium ion, made by the reaction between an ammonia molecule and a hydrogen ion, can be represented as shown in the diagram below.

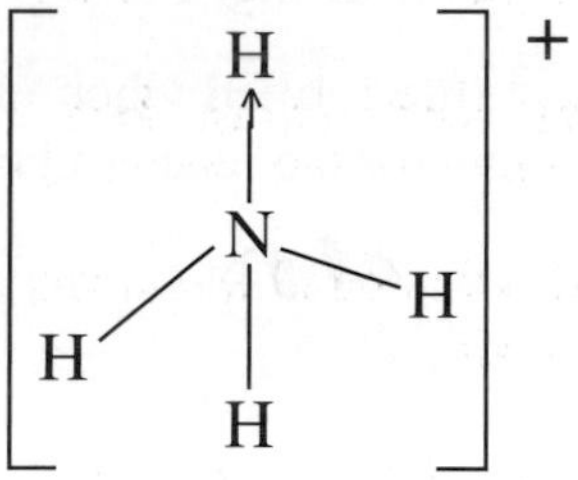

(i) Name the type of bond represented in the diagram by N–H.

It represents a covalent bond. ✓

(ii) Name the type of bond represented in the diagram by N→H.

It represents a dative ^ bond.

(iii) In terms of electrons, explain why an arrow is used to represent this N→H bond.

A ^ pair of electrons on the nitrogen is given to form a bond ✓ with the hydrogen atom.

(iv) In terms of electron pairs, explain why the bond angles in the NH_4^+ ion are all 109° 28′.

Between each hydrogen atom and the nitrogen there is a shared pair of electrons. ✓ The four hydrogen atoms are all δ+ and so repel each other. ✗ The shape with the minimum repulsion is a tetrahedron ✓ which has an internal angle of 109° 28′.

[7 marks] 4/7

(b) Define the term *electronegativity*.

Electronegativity is the extent to which an element pulls a pair ✓ of electrons ^ to itself.

[2 marks] 1/2

(c) A bond between nitrogen and hydrogen can be represented as:

$$\overset{\delta-}{N}—\overset{\delta+}{H}$$

(i) In this representation, what is the meaning of the symbol δ+?

The symbol δ+ means that the atom has gained a positive charge.

(ii) From the bond representation, what can be deduced about the electronegativity of hydrogen relative to that of nitrogen?

The nitrogen has drawn electrons to itself and so hydrogen is less electronegative than nitrogen. ✓

[2 marks] 1/2

6/11 [Total 11 marks]

2 Sodium metal, sodium fluoride and hydrogen fluoride all form different types of solids with different bonding.

(a) State the type of solid in sodium, sodium fluoride and hydrogen fluoride.

Sodium is metallic, ✓ sodium fluoride ionic ✓ and hydrogen fluoride is molecular.

[3 marks] 2/3

(b) Describe the bonding in metallic sodium and hence explain why it conducts electricity in the solid state.

Each sodium atom loses its s electron which form a sea of delocalised ✓ electrons which are free ✓ to move and so the solid conducts electricity.

[4 marks] 2/4

(c) Describe the bonding in sodium fluoride and explain why it conducts electricity when molten but not when solid.

A sodium atom gives ✗ an electron to a fluorine atom so that both reach the stable ✗ electronic structure of a noble gas.

In the liquid the ions are free to move ✓ around.

[4 marks] 1/4

5/11 [Total 11 marks]

How to score full marks

1 (a) (ii) The correct name for this type of bond is a **dative covalent** bond.

(iii) It is the **lone** pair of electrons on the nitrogen atom that is given to form the bond.

(iv) The theory of shapes is called the Valence Shell Electron Pair Repulsion (VSEPR) theory. It is the **electron pairs** around the nitrogen atom that repel, not the hydrogen atoms. This is a common error, which spoilt an otherwise excellent answer.

(b) The definition is about pulling a pair of electrons which are in a **covalent bond**.

(c) (i) The symbol δ+ means a **partial** positive charge; one that is less than a 1+ ionic charge.

2 (a) The correct answer is that hydrogen fluoride forms a **hydrogen bonded molecular solid**. A mark is lost here because the phrase 'hydrogen fluoride is molecular' is ambiguous. **There are two types of molecular solid: simple and hydrogen bonded molecular**.

(b) To score full marks you must mention the fact that the positive ions form a **regular lattice** and that the bond is the **electrostatic attraction** between the Na^+ ions and the delocalised electrons.

(c) At this level, it is correct to describe an ionic bond as **the force of attraction between oppositely charged Na^+ and F^- ions** that results from the **loss** of an electron from the sodium atom and the **gain** of an electron by the fluorine atom.

The question is in two parts and the student ignored some of the second part, and did not explain why the solid does not conduct. The correct answer is that the ions in **solid** sodium fluoride are fixed in position and are **not free to move**, and so **cannot conduct** an electric current.

Don't make these mistakes ...

In your explanation of an ionic bond do **not** write about noble gas structures. An ionic bond is the **attraction** between **oppositely** charged ions formed by the transfer of electrons.

Do not state that covalent bonds are weak. They are about the **same** strength as ionic bonds. The weak forces/bonds are intermolecular forces – those **between** molecules.

Do not confuse intermolecular forces (hydrogen bonds and van der Waals forces) with covalent bonding. The former are **between** molecules and the latter are **within** molecules.

Do not forget to count **lone pairs** when deducing shapes of molecules or ions. The number of lone pairs can be calculated using the formula:

Number of lone pairs

$$= \frac{\text{(group number of the atom – the number of covalent bonds)}}{2}$$

But for a negative ion add 1 electron per negative charge to the group number.

For a positive ion subtract 1 electron per positive charge from the group number.

When deducing bond angles, do not forget **lone pairs**. Thus, ammonia, which has **one lone pair**, is pyramidal with a bond angle of less than 109.5° and water, which has **two lone pairs**, is V shaped with a bond angle less than that of ammonia.

Key points to remember

Chemical bonds

An **ionic bond** is the electrostatic attraction between a positive ion (the cation) and a negative ion (the anion). The strength depends on two factors:

- The charge on the ions – the larger the stronger the bond
- The sum of the ionic radii – the smaller the stronger the bond.

Cations are formed by the **loss** of one or more electrons. e.g. Na (2,8,1) forms Na^+ (2,8).

Anions are formed by the **gain** of one or more electrons. e.g. Cl (2,8,7) forms Cl^- (2.8.8).

Polarisation

The electrons on an **anion** can become polarised and this causes the bond to have some covalent character.

LiI is partially covalent, whereas LiF is totally ionic. A cation that is **small** and/or is **highly charged** has a large polarising power. e.g. the doubly charged Mg^{2+} ion is more polarising than the singly charged Na^+ ion.

An anion that is **large** and/or is **highly charged** is easily polarised. e.g. the large I^- is more easily polarised than the smaller F^- ion.

A **covalent** bond is the **sharing of a pair** of electrons by two atoms.

It is formed when atomic orbitals, each containing **one** electron, **overlap**.

A **short** covalent bond is **stronger** than a longer bond, e.g. the C—Cl bond is stronger than the C—I bond.

A π bond is weaker than a σ bond.

The C═C bond strength is less than twice the C—C bond strength.

A σ bond is formed by the 'head on' overlap of 2 orbitals, usually two p orbitals. The result is an electron cloud that is on a line between the nuclei:

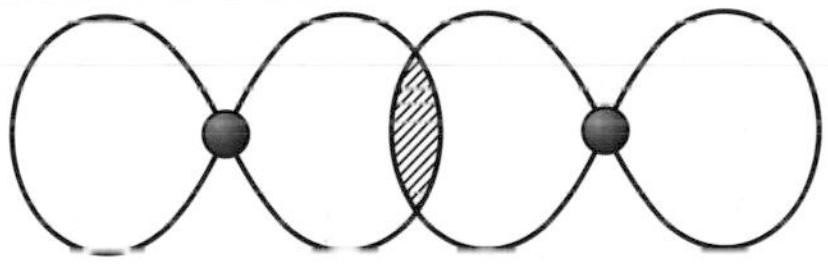

A π bond is formed by the 'sideways' overlap of two p orbitals, resulting in the electron cloud being above and below the line between the nuclei:

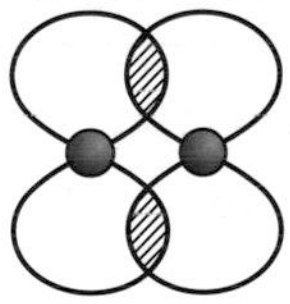

A **dative covalent (co-ordinate)** bond is formed when one of the atoms provides **both** electrons to the shared pair.

The result is a shared pair of electrons as in an ordinary covalent bond.

The ammonium ion contains one dative and three covalent bonds.

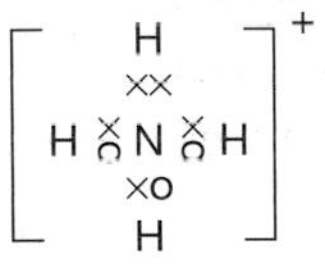

A **polar** covalent bond results if the electronegativities of the two atoms are different.

The electronegativity of an element is defined as the extent to which it draws the electrons in a covalent bond towards itself.

The electronegativities increase across a period and decrease down a group. Thus fluorine is the most electronegative element.

HCl and HBr are both polar molecules, but in totally symmetrical molecules, such as CH_4 and CO_2, the polarity of each bond cancels out and the molecule is itself not polar.

Key points to remember

A **metallic** bond occurs when the metal atoms lose their outer electrons. These then form a 'sea' of delocalised electrons in a regular lattice of positive ions.

The bond is the force of attraction between the delocalised electrons and the positive ions.

The electrons in a metal are able to move through the lattice and so the solid conducts electricity.

Types of solid		**Examples**
Ionic		NaCl, MgO, $Ba(OH)_2$
Metallic		All of groups 1 and 2 and aluminium, tin and lead and the *d*-block elements
Molecular:	simple	I_2, CO_2
	hydrogen bonded	HF, H_2O, NH_3, alcohols
Giant atomic (or giant covalent or giant molecular)		Graphite, diamond, silicon and quartz (SiO_2)

Intermolecular forces

The bonds **between** covalent molecules are called **intermolecular** forces and are much weaker than the strong covalent bonds **within** the molecule.

Hydrogen bonds

A hydrogen bond occurs between a δ+ H atom in one molecule and a δ− F, O or N atom in another molecule.The hydrogen bonds in ice are the dotted lines and the covalent bonds are the solid lines:

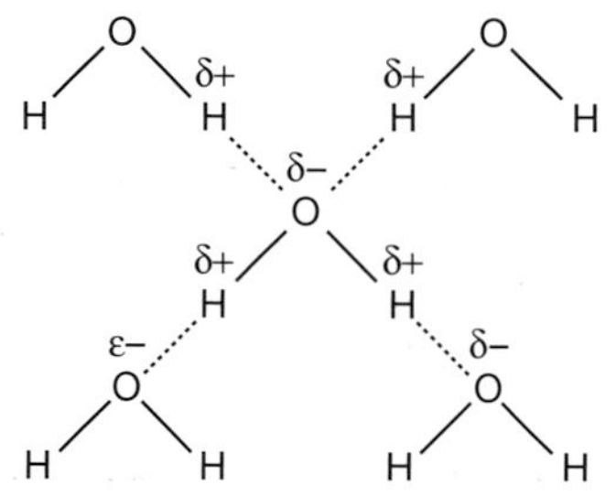

Change of state

Heating a solid until it just liquefies.

The **stronger** the intermolecular force the **higher** the melting point.

As a solid is heated the particles **vibrate more** until the energy is sufficient to **break up** the lattice, the solid melts and the particles become **free to move** about.

Instantaneous induced dipole/induced dipole (or dispersion force)

This is significant for larger molecules. The strength depends mainly on the total **number of electrons** in the molecule. This dispersion force in HBr is stronger than the dispersion force in HCl because HBr has 36 electrons and HCl has only 18.

HCl has a higher boiling point than Ar even though the dispersion forces are about the same strength, because HCl is also polar.

Permanent dipole/dipole force

This is significant if the molecule is small and very polar.

A good example is HCl, which is very polar:

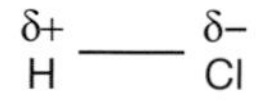

Heating a liquid until it boils

The **stronger** the intermolecular force the **higher** the boiling point.

As a liquid is heated, its particles gain kinetic energy and are more likely to break free from the forces holding them in the liquid. They therefore escape from the surface by evaporation.

As the boiling point is reached, the particles gain sufficient energy to escape from the inside of the liquid.

Shapes of molecules and ions

The shape of a molecule or ion is determined by the number of **bond** pairs and **lone** pairs of electrons around the central atom. The electron pairs repel each other and get as far apart as possible. The lone pair/ lone pair repulsion is greater than lone pair / bond pair, which is greater than bond pair / bond pair.

l.p/l.p > l.p/b.p > b.p/b.p.

The bond angles depend on the arrangement of electrons and the number of lone pairs. In a tetrahedral molecule it is 109.5°. However, a lone pair reduces that angle, and two lone pairs reduce it further. The following table shows some bond angles:

Numbers of pairs	Shape	Example	Bond angle
2 bond pairs	linear	$BeCl_2$ 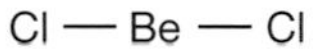	180°
3 bond pairs	triangular	$AlCl_3$ 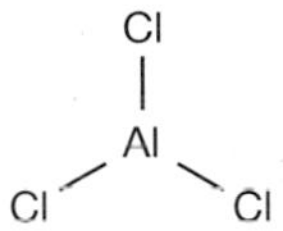	120°
4 bond pairs	tetrahedral	CH_4 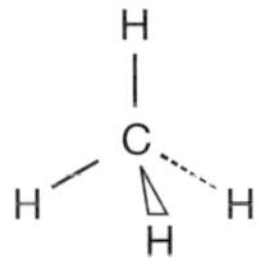	109.5°
3 bond + 1 lone pair	pyramidal	NH_3 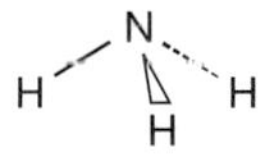	107°
2 bond + 2 lone pairs	bent (non-linear)	H_2O 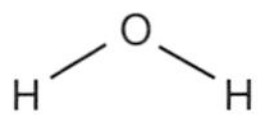	104.5°
5 bond pairs	triangular bipyramid	PCl_5 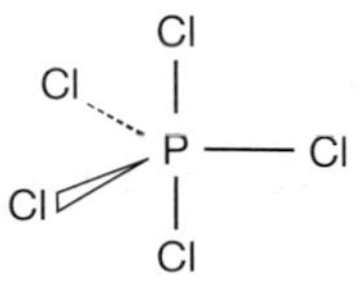	120° and 90°
6 bond pairs	octahedral	SF_6 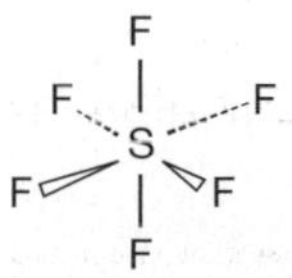	90° and 180°

Questions to try

Examiner's hints ***for question 1***

First state the theory in general terms.
Make the point about bond and lone pairs and their effect on shape and angle.
Then give examples of four different shapes, one of which must contain one or more lone pairs.
Draw diagrams of the shapes and mark in the bond angles.

Q1

Explain how you would predict the shapes of, and the bond angles in, simple molecules. In your answer, you should choose examples of **four** different molecular shapes.

....

....

....

....

.... [10 marks]

[Total 10 marks]

Examiner's hints ***for question 2***

(a) Hydrogen bonds are drawn as dotted lines and covalent bonds as solid lines.
(b) First identify the type of intermolecular forces and then state their relative strengths. Don't forget to compare HF with HCl and then HCl with HBr.

Q2

(a) What is meant by the term hydrogen bond? Illustrate your answer by reference to the bonding in water.

....

.... [4 marks]

(b) Explain why HF has a higher boiling temperature than HCl, but HCl has a lower boiling temperature than HBr.

....

.... [4 marks]

[Total 8 marks]

Examiner's hints ***for question 3***

(a) Don't forget that boiling is about the separation of the molecules, not the breaking of covalent bonds.
(b) In describing bonding, you must state where the electrons come from and where they end up.
(c) Count the number of bond and lone pairs. A 1+ ion has lost an electron. Make sure that your diagram shows the 3-dimensional shape of the substance.

Q3

(a) Both HF and HCl are molecules having a polar covalent bond. Their boiling points are 293K and 188K respectively.

(i) State which property of the atoms involved causes a bond to be polar.

....

(ii) Explain in terms of the intermolecular forces present in each compound, why HF has a higher boiling point than HCl.

....

.... [5 marks]

(b) When aluminium chloride reacts with chloride ions, as shown in the equation below, a co-ordinate (dative covalent) bond is formed.

$$AlCl_3 + Cl^- \rightarrow AlCl_4^-$$

Explain how this co-ordinate bond is formed.

....

.... [2 marks]

(c) Draw the shape of the PCl_5 molecule and of the PCl_4^+ ion. State the value(s) of the bond angles.

PCl_5 PCl_4^+

Bond angle(s) .... Bond angle(s) ....

[5 marks]

[Total 12 marks]

Examiner's hints ***for question 4***

(a) Don't forget to describe the forces that hold the solid together. Conduction of electricity is a flow of charged particles. Which particles move in this example?

(b) What are the particles in the solid and what holds them together? Why cannot charged particles move?

(c) What are the particles? What are they doing at room temperature and in what way does this change as the solid begins to be heated? What happens at the melting point and what particles can now move and conduct electricity?

Q4

Sodium and sodium chloride are solids at room temperature.

(a) Describe the bonding in solid sodium metal, and hence explain why it conducts electricity. [3 marks]

(b) Explain the bonding in solid sodium chloride and explain why it does not conduct electricity. [3 marks]

(c) **(i)** Describe the change in the motion and position of the particles as solid sodium chloride is heated from room temperature to just above its melting point.

(ii) Explain why molten sodium chloride can conduct electricity. [4 marks]

[Total 10 marks]

The answers to these questions are on pages 80–82.

Exam Question and Student's Answer

1 **(a)** The periodic table is divided into groups and periods. What is the difference in electronic configuration between:

(i) two elements one above the other in a group?

They have the same number of electrons in their outer shells ✗

[1 mark] 0/1

(ii) two adjacent elements in a period?

One has one more electron than the other. ✓

[1 mark] 1/1

(b) Explain, in terms of structure and bonding, why lithium and graphite conduct electricity in the solid state.

Lithium is a metal. Its outer 2s electron is part of a delocalised ✓ cloud of electrons between a regular arrangement of positive Li^+ ions. These electrons can move through the solid. ✓

In graphite, the π electrons are in a delocalised ✓ cloud above and below the plane ✓ of the molecule. They also can move and so it conducts electricity.

[5 marks] 4/5

(c) The melting points of the elements in period 3 are shown below:

	sodium	magnesium	aluminium	silicon	phosphorus	sulphur	chlorine	argon
m.pt/°C	98	650	660	1410	44	119	– 101	– 189

Explain why the melting point of:

(i) aluminium is greater than that of sodium

Aluminium has three delocalised electrons, and so the forces between the delocalised electrons and the Al^{3+} ion is greater than between the delocalised electrons and the Na^{+} ion.

[3 marks] 2/3

(ii) silicon is the highest in the group

The melting point reaches a maximum at silicon because it can form 4 bonds. Sodium and chlorine form one, magnesium and sulphur two, aluminium and phosphorus three, so the bonding in silicon is the strongest. ✗

[4 marks] 0/4

(iii) sulphur is higher than that of phosphorus.

The bonding in sulphur is stronger because, being smaller, it forms stronger covalent bonds.

[3 marks] 0/3

(d) Some data is shown below about silicon:

electronegativity	1.8
atomic radius / nm	0.117
1st ionisation energy / kJ mol^{-1}	786

State and explain which of aluminium or phosphorus has:

(i) an electronegativity of 2.1

Phosphorus, because the electronegativity increases left to right in the periodic table as the atoms become smaller, and so the nucleus has a greater pull on the bonding electrons.

[2 marks] 2/2

(ii) an atomic radius of 0.110 nm

Aluminium, as the atoms get bigger as they have more electrons.

[2 marks] 0/2

(iii) a 1st ionisation energy of 577 kJ mol^{-1}.

Aluminium, as the 1st ionisation energies increase across a period as the nuclear charge increases without an increase in the number of inner (shielding) electrons.

[2 marks] 2/2

[Total 23 marks]

11/23

How to score full marks

1 (a) (i) **The student does not answer the question**, which asks for the **difference** not the similarity. **Take care to read the question carefully**. The correct answer is:

The element higher in the group has **one fewer electron shell** than the one lower in the group.

(b) The explanation of why lithium conducts electricity is correct, but the answer for graphite is incomplete. The correct answer is:

The structure of graphite consists of **interlocking hexagons** in a plane. One electron from each carbon atom forms a **delocalised (π) cloud above and below** the plane. **These delocalised electrons can also move through the solid and so graphite conducts electricity**.

(c) (i) The student omitted one important reason why aluminium has the higher melting point. He or she should have also written:

The Al^{3+} ion is much smaller than the Na^+ ion. This means that the metallic bond is much stronger and so the melting point is higher.

(ii) The answer here is totally wrong. The correct answer is:

Silicon forms a **giant atomic (covalent or molecular)** lattice. All the **covalent bonds** must be broken before silicon will melt. Covalent bonds are much **stronger than intermolecular** forces in the other non-metals, and **stronger than the metallic** bond in the s and p block metals, so silicon has the highest melting point.

(iii) This answer is also totally wrong. The correct answer is:

Sulphur forms S_8 molecules and phosphorus forms P_4 molecules. There **are more electrons** in a sulphur molecule than in a phosphorus molecule, so the **intermolecular** forces are stronger, hence sulphur has a higher melting point.

Note: when a molecular solid melts, it is the **intermolecular** forces that are broken, **not** the covalent bonds between atoms **within** the molecule.

(d) (ii) The correct answer is phosphorus, as the atoms get **smaller** across a period because the nuclear charge **increases**.

Don't make these mistakes ...

- Don't think that the atoms get larger across a period. They get **smaller** because, as the nuclear charge increases, the electrons get pulled in more.
- Don't get the formulae of **group II compounds** wrong:
  Oxides are MO
  Hydroxides are $M(OH)_2$
  Carbonates are MCO_3
  Sulphates are MSO_4
  Nitrates are $M(NO_3)_2$
- Don't forget that the solubility of **sulphates decreases** down group II and the solubility of **hydroxides increases** down group II.
- Don't confuse **intermolecular** forces, which are **weak**, with **covalent** bonds, which are **strong**.
- Don't forget to learn the formulae of the **non-metallic elements** in periods 2 and 3.
- Don't forget to check that the formulae are correct when writing an equation. Check also that the **equation balances**.

Key points to remember

The Periodic Table

- Elements are arranged according to **atomic number**.
- The periodic table is divided into horizontal rows called **periods**.
- It is divided into vertical columns called **groups**.
- The **s** block consists of groups I and II.
- The **p** block consists of groups III, IV, V, VI, VII and the noble gases (group 0).
- The **d** block is the elements scandium to zinc and those below them.

Trends across a period

Atomic radius	This **decreases** left to right because the nuclear charge increases without an increase in inner (shielding) electrons.
Electronegativity	This **increases** left to right because the smaller atoms exert a greater pull on the bonding electrons.
1st ionisation energy	The general trend is to **increase** left to right as the nuclear charge increases without any increase in shielding.
Electrical conductivity	**Metals** (and **graphite**) conduct; non-metals do not conduct.

Melting and boiling points

To explain the variation in melting and boiling points, you must:

- Decide on the **type of solid** (metallic, giant atomic/covalent/molecular or simple molecular).
- Decide on the **type of force** (due to delocalised electrons, covalent bond or van der Waals force).
- Comment on the **strength of the force** (covalent bond > metallic bond in s and p block > van der Waals force).

Strength of force depends on various factors:

- If metallic, the strength depends on the **number of delocalised electrons and the size** of the metal ion (the smaller the ion, the higher the melting point).
- If giant atomic/covalent/molecular, the strength depends on the **length of the covalent bonds**. Shorter bonds are stronger; the C–C bond length is less than the Si-Si and so the C–C bond is stronger.
- If molecular, the strength depends on the **number of electrons in the molecule**. Therefore the melting (and boiling) points are in the order:

$S_8 > P_4 > Cl_2 > Ar$

$F_2 > O_2 > N_2 > Ne$

Trends in Group II (Group 2) of the Periodic Table

Atomic radius	**Increases** down a group as the number of shells of electrons increases.
Electronegativity	**Decreases** down a group as the atoms become larger.
1st ionisation energy	**Decreases** down a group as the outer electron becomes further away from the nucleus. More shielding occurs with increase in nuclear charge.
Reactivity	Reactivity as a metal **increases** down the group because it becomes easier to remove two electrons from the outer shell.

Reactions of Group II elements and compounds

Reaction of element (M) with oxygen	All burn to form an oxide: $2M + O_2 \rightarrow 2MO$
Reaction of element with water	Be does not react. Mg burns in steam: $Mg + H_2O \rightarrow MgO + H_2$ The others react rapidly with cold water: $Ca + 2H_2O \rightarrow Ca(OH)_2 + H_2$
Reaction of oxide (MO) with dilute hydrochloric acid	All react as bases: $MO + 2HCl \rightarrow MCl_2 + H_2O$
Reaction of carbonate (MCO_3) with dilute hydrochloric acid	All react to form carbon dioxide gas: $MCO_3 + 2HCl \rightarrow MCl_2 + CO_2 + H_2O$
Effect of heat on carbonate	$BaCO_3$ is stable to heat. The stability **increases** down the group. The others decompose: $CaCO_3 \rightarrow CaO + CO_2$
Solubility of sulphate (MSO_4)	The solubility **decreases** down the group. $BaSO_4$ is insoluble and is the precipitate in the test for sulphates.
Solubility of hydroxide ($M(OH)_2$)	The solubility **increases** down the group. $Mg(OH)_2$ is almost insoluble, $Ca(OH)_2$ slightly soluble and $Ba(OH)_2$ fairly soluble.

Questions to try

Examiner's hints *for question 1*

(a) (i) First you must identify the type of solid and therefore the forces between the particles and their strengths. For molecular substances you must write down the molecular formula.

(ii) For explanations of ionisation energies it is essential to know which electron in which orbital is being removed.

(b) Remember that water can be solid, liquid or gas.

Q1

(a) The table below gives the melting points and the 1st ionisation energies for the elements in Period 3 of the Periodic Table.

Element	Na	Mg	Al	Si	P	S	Cl	Ar
Melting point / K	371	923	933	1680	317	392	172	84
1st ionisation energy / kJ mol^{-1}	494	736	577	786	1060	1000	1260	1520

(i) In terms of structure and bonding, explain why silicon has a very high melting point, and why the melting point of sulphur is higher than that of chlorine.

....

....

....

....

....

....

[5 marks]

(ii) Write the electronic configurations, using the $1s^2$ notation, of phosphorus and sulphur and hence explain why the 1st ionisation energy of phosphorus is higher than that of sulphur.

....

....

....

....

[2 marks]

(b) Give the conditions under which magnesium and calcium react rapidly with water. For each reaction, state **one** observation that you would make and write the equations for the reactions.

....

....

....

....

....

[4 marks]

[Total 11 marks]

Examiner's hints ***for question 2***

(a) The explanation for trends depends on:

- The nuclear charge.
- The number of inner (shielding) electrons.
- The radius of the atom or ion.

(b) Metals form positive ions when they react.

Q2

Use the data below to answer the questions that follow:

	Be	Mg	Ca	Sr	Ba
1st ionisation energy / kJ mol^{-1}	900	736	590	548	502
2nd ionisation energy / kJ mol^{-1}	1760	1450	1150	1060	966
Metallic radius / pm	112	160	197	215	217

(a) Explain the trends in 1st ionisation energy.

....

....

....

....

[3 marks]

(b) Magnesium reacts very slowly with hot water, but burns when heated in steam.

Calcium reacts rapidly with cold water.

Barium reacts even more rapidly with cold water.

(i) Write the equation (with state symbols) for calcium's reaction with water.

.... [2 marks]

(ii) Write the equation (with state symbols) magnesium's reaction with steam.

.... [1 mark]

(iii) Explain the trend in reactivity of the group II metals with water.

....

.... [2 marks]

(c) Radium is in group II and is below barium. Predict:

(i) the reaction between radium and water

....

.... [2 marks]

(ii) the solubility of radium sulphate

.... [1 mark]

(iii) the metallic radius of radium.

.... [1 mark]

[Total 12 marks]

The answers to these questions are on pages 82 and 83.

Exam Question and Student's Answer

1 (a) Deduce the oxidation number of iodine in the following species.

(i) I_2O_7

Oxygen is −2, so 7 oxygen atoms are −14, so each iodine is +7. ✓

(ii) IO_4^-

4 oxygen atoms are −8, so iodine is +8. ✗

[2 marks] 1/2

(b) Iodine, I_2, can be reduced to iodide ions, I^-, by tin(II) ions, Sn^{2+}, which are themselves oxidised to tin(IV) ions, Sn^{4+}.

(i) Construct the oxidation and reduction half-equations for the above system.

$Sn^{2+} \rightarrow Sn^{4+} + 2e^-$ ✓

$I_2 + 2e^- \rightarrow 2I^-$ ✓

(ii) Use the above half-equations to construct the overall ionic equation for the reaction.

$Sn^{2+} + I_2 + 2e^- \rightarrow Sn^{4+} + 2e^- + 2I^-$ ✗

[3 marks] 2/3

(c) When chlorine, Cl_2, is added to aqueous sodium bromide, NaBr, a reaction takes place.

(i) State what you would see in this reaction.

It goes ^ red-brown.

(ii) Write an equation for this reaction.

Cl_2 + ^ NaBr → ^ NaCl + Br_2

(iii) What happens to electrons during this reaction?

The NaBr ✗ loses an electron and the chlorine atom in Cl_2 gains it. ✓

(iv) Why does no reaction take place when bromine is added to aqueous sodium chloride?

Because bromine is a weaker oxidising agent than chlorine. ✓

(v) Describe a simple test to confirm the presence of iodide ions in aqueous sodium iodide.

Add aqueous silver nitrate to the solution. A yellow precipitate ✓
is obtained which is insoluble in concentrated aqueous ammonia. ✓

[8 marks] 4/8

(d) Concentrated sulphuric acid can be reduced by some solid sodium halides to H_2S.

(i) Give **one** solid sodium halide which will reduce concentrated sulphuric acid, forming H_2S.

Sodium iodide ✓ will reduce sulphuric acid to H_2S.

(ii) Write a half-equation for the formation of H_2S from sulphuric acid.

$H_2SO_4 + e^- \rightarrow 2O_2 + H_2S$ ✗

[2 marks] 1/2

(e) A different solid sodium halide reacts with concentrated sulphuric acid without reduction forming a halogen containing product **X**.

(i) Suggest an identity for **X**.

Hydrochloric acid ✗

(ii) Identify the solid sodium halide which produces **X**.

Sodium chloride ✓

(iii) State the role of sulphuric acid in the formation of **X**.

It removes the Na from the NaCl. ✗

(iv) Write an equation for the reaction with concentrated sulphuric acid in which **X** is formed.

$NaCl + H_2SO_4 \rightarrow HCl + NaHSO_4$ ✓

[4 marks] 2/4

10/19 [Total 19 marks]

How to score full marks

(a) (ii) The student forgot that the sum of the oxidation numbers in IO_4^- adds up to –1, which is **the charge on the ion**. So iodine is in the +7 state (+7 –8 = –1).

(b) (ii) The electrons must be cancelled in the overall equation formed **by adding the two half-equations**.

(c) (i) Observations about the colour change in a solution **must state the colour before as well as after**. The correct answer is 'the colourless solution turns red-brown'.

(ii) The species are correct but **the equation does not balance**. Always check that your equations balance. Here you need 2NaBr on the left and 2NaCl on the right.

(iii) It is the **Br^- ions** not the NaBr that lose electrons.

(v) **Dilute nitric acid** must also be added in the halide test. This is to prevent other precipitates (of carbonates or hydroxides) forming.

(d) (ii) This is a very difficult half-equation. First you must realise that only the oxidation number of S decreases as H_2SO_4 is reduced. So O_2 cannot be a product as that would require the oxygen atoms in the H_2SO_4 being oxidised. H^+ ions must be added on the left so that H_2O can be formed on the right. **The number of electrons equals the change in oxidation number**. Here the sulphur goes from +6 in H_2SO_4 to –2 in H_2S, a change of 8. The half-equation is:

$$H_2SO_4 + 8H^+ + 8e^- \rightarrow H_2S + 4H_2O$$

Note that the equation also balances for charge.

(e) (i) Hydrochloric acid is the name for the aqueous solution. The gas is called **hydrogen chloride**.

(iii) The sulphuric acid protonates the Cl^- ions in the NaCl, and so is acting as an **acid**.

Don't make these mistakes ...

Don't forget OIL RIG – **O**xidation **I**s **L**oss and **R**eduction **I**s **G**ain

Don't get muddled about redox. Remember that **oxidising agents** (such as chlorine) become **reduced**, and **reducing agents** (such as Fe^{2+} ions and iodide ions) become **oxidised**.

Don't forget that ionic half equations contain electrons. **Make sure that you get the electrons on the correct side**. When a substance is reduced, electrons are gained and so go on the left-hand side of a half equation, and when a substance is oxidised, electrons are lost and so are on the right in a half equation.

Don't forget that the number of electrons in a half equation equals the total change in oxidation number. If the substance is in acid solution you will probably need to have H^+ on the left and H_2O on the right.

Key points to remember

REDOX

Oxidation number or state is the charge that an element would have if all the bonds in the substance were ionic.

- Uncombined elements are zero.
- F is always −1, Group I are +1, Group II +2, oxygen is −2 (except in peroxides), H is +1 (except in metallic hydrides).
- Oxidation numbers in a neutral compound add up to zero and in a polyatomic ion they add up to the charge on that ion.

  Some examples:
- Mn in MnO_4^- is +7 as each O is −2 = −8, so Mn + (−8) = −1, therefore Mn is +7.
- Cr in $Cr_2O_7^{2-}$ is +6 as each O is −2 = −14, so 2Cr + (−14) = −2, therefore each Cr is +6.
- Cl in ClO^- is +1 as O is −2 and so Cl + (−2) = −1, therefore Cl is +1.
- O in H_2O_2 (hydrogen peroxide) is −1, as each H is +1, so the two O atoms between them are −2.

Half equations for redox processes contain electrons.

- If a species is **reduced**, the electrons are on the **left**-hand side of the half equation.
- If a species is **oxidised**, the electrons are on the **right**-hand side of the half equation.
- The number of electrons is equal to the total change in oxidation number.
- For oxidising agents that contain oxygen (such as MnO_4^-), you will need H^+ on the left and H_2O on the right.

  Some examples of half equations:
- for chlorine being reduced: $Cl_2 + 2e^- \rightarrow 2Cl^-$
- for manganate(VII) ions being reduced:

  $$MnO_4^- + 8H^+ + 5e^- \rightarrow Mn^{2+} + 4H_2O$$

  (The oxidation number of Mn changes by −5, so there have to be 5 electrons on the left.)
- for iodide ions being oxidised: $2I^- \rightarrow I_2 + 2e^-$

- **O**xidation **I**s **L**oss of electrons
- **R**eduction **I**s **G**ain of electrons
- Remember **OIL RIG**.

Overall redox equations can be obtained by adding half equations. You must check that:

- The number of electrons on the left of one half equation equals the number of electrons on the right of the other half equation.
- In the overall equation, the total change in oxidation number of the substance being reduced equals the total change in the substance being oxidised.
- The two reactants are on the left of the equation.

The overall equation for the oxidation of I^- ions by MnO_4^- ions is obtained from the two half equations:

$$MnO_4^- + 8H^+ + 5e^- \rightarrow Mn^{2+} + 4H_2O \quad \text{and} \quad 2I^- \rightarrow I_2 + 2e^-$$

The MnO_4^- half equation has 5 electrons, but the I^- equation only 2, so the MnO_4^- equation has to be multiplied by 2 and the I^- equation by 5, so that both have 10 electrons. They are then added to give

$$2MnO_4^- + 16H^+ + 10I^- \rightarrow 2Mn^{2+} + 8H_2O + 5I_2$$

Group VII (Group 7)

The halogens

The boiling points of the halogens increase down the group.

- Fluorine and chlorine are gases, bromine is a volatile liquid, iodine is a solid. This is because the number of electrons in the halogen molecule increases, causing an increase in the strength of the intermolecular forces.

The oxidising power of the halogens decreases down the group.

- Chlorine will oxidise bromide ions but bromine will not oxidise chloride ions. Chlorine and bromine will oxidise Fe^{2+} ions but iodine will not. This is because the smaller chlorine atom has a greater attraction for electrons than the other halogens which are larger.

Chlorine (and bromine) reacts with water:

$$Cl_2 + H_2O \rightleftharpoons HCl + HOCl$$

Chlorine (and bromine) reacts with alkalis such as sodium hydroxide solution:

$$Cl_2 + 2OH^- \rightarrow Cl^- + OCl^- + H_2O$$

- In both reactions the chlorine has disproportionated, because it has been both oxidised (oxidation number goes up from 0 in Cl to +1 in HOCl) and reduced (oxidation number goes down from 0 in Cl to −1 in HCl).

Test for the halogens

- Chlorine is very pale green and bleaches damp litmus. It also turns colourless potassium bromide solution red-brown.
- Bromine is red-brown and slowly bleaches damp litmus.
- Iodine turns starch blue-black and solutions of iodine in hexane are violet.

The halides

Test for ionic halides (in solution):

Add dilute nitric acid and aqueous silver nitrate. Then add dilute ammonia, followed by concentrated ammonia to the precipitate that forms (this is a silver halide).

- Chlorides give a white ppt, bromides a pale cream ppt and iodides a pale yellow ppt
- The ppt of AgCl dissolves in dilute ammonia. The ppt of AgBr dissolves in conc ammonia but not dilute ammonia. The ppt of AgI does not dissolve.

Reducing power of halides

- Iodide ions are the strongest reducing agents and will reduce Fe^{3+} ions to Fe^{2+}.
- Bromide ions are a much weaker reducing agent, and chloride ions weaker still. Neither will reduce Fe^{3+} ions.
- When concentrated sulphuric acid is added to a solid ionic halide, gaseous hydrogen halide is first produced. This may then be oxidised to the halogen by the conc sulphuric acid, depending on the strength of the hydrogen halide as a reducing agent.
- When conc sulphuric acid is added to a chloride, steamy fumes of HCl are produced. This is too weak a reducing agent to be oxidised to chlorine by the sulphuric acid.
- When conc sulphuric acid is added to a solid bromide, misty fumes of HBr are produced some of which is oxidised by the sulphuric acid to red-brown bromine vapour. Some SO_2 is made by the reduction of the H_2SO_4.
- Hydrogen iodide is rapidly oxidised to violet fumes of iodine and reduces H_2SO_4 to H_2S and S.

Hydrogen halides as acids

- HCl, HBr and HI are all strong acids, but they are not as strong as sulphuric acid:

$$HCl(aq) + H_2O(l) \rightarrow H_3O^+(aq) + Cl^-(aq)$$

- Concentrated sulphuric acid will protonate halide ions to form the hydrogen halide, but see the box on 'Reducing power of halides' (left) for subsequent reactions.

Questions to try

Examiner's hints *for question 1*

(a) (i) Remember that the oxidation numbers in a polyatomic ion add up to the charge on the ion, and in a neutral substance they add up to zero.

(c) (i) An oxidising agent is reduced, and so it gains electrons. As the oxidising agent, FeO_4^-, is in acid solution, you will need to have H^+ ions on the left of the half equation. Remember that the number of electrons equals the change in oxidation number.

Q1

(a) (i) Define oxidation number (oxidation state).

....

....

[2 marks]

(ii) What is the oxidation number of chlorine in:

Cl_2 .... OCl^- .... $NaClO_3$ ....

[3 marks]

(b) (i) Write balanced equations for the reactions of chlorine with:

water .... [1 mark]

hydroxide ions .... [1 mark]

(ii) What, in terms of oxidation and reduction, has happened to the chlorine in these reactions?

....

....

[2 marks]

(c) Iron forms a highly oxidising ion of formula FeO_4^-.

(i) Write the half equation for its reduction to Fe^{3+} ions in acid solution.

....

[2 marks]

(ii) Write the half equation for the oxidation of Fe^{2+} ions to Fe^{3+} ions.

....

[1 mark]

(iii) Hence write the overall equation for the reaction between FeO_4^- ions and Fe^{2+} ions.

....

[2 marks]

[Total 14 marks]

Examiner's hints ***for question 2***

(a) First identify the structure (metallic, giant atomic, molecular), then the type of force that needs to be overcome to separate the particles. Lastly compare the strength of the force and hence the energy required for separation down the Group.

(b) You see bubbles, not a gas; a precipitate or a colour change, not the identity of the chemical produced.

(c) Check that you know the colours of the halogens and remember that all halide ions are colourless in solution.

Q2

(a) Describe and explain the trend in boiling points of the elements down Group VII from fluorine to iodine.

....

....

....

[4 marks]

(b) Describe what you would observe when aqueous silver nitrate, followed by dilute aqueous ammonia, is added to separate aqueous solutions of sodium chloride and sodium bromide.

....

....

....

[4 marks]

(c) State the trend in oxidising abilities of the elements down Group VII from chlorine to iodine.
Explain how this trend can be shown by displacement reactions between halogens and halide ions in aqueous solutions.
Illustrate your answer with appropriate observations and equations.

....

....

....

....

[7 marks]

[Total 15 marks]

The answers to these questions are on pages 83–85.

Exam Questions and Student's Answers

1 (a) This part of the question is about the chemical reaction between the two gases, oxygen, O_2, and nitrogen monoxide, NO.

$$2NO(g) + O_2(g) \rightarrow 2NO_2(g)$$

(i) Define the term **standard enthalpy of formation, $\Delta H_f^\ominus$**.

The standard enthalpy of formation is the heat energy change when the substance is formed from its elements under standard conditions of 1 atm pressure and at 298K. ✓

(ii) Use the data below to calculate the enthalpy change for the reaction between oxygen and nitrogen monoxide.

$$\Delta H_f^\ominus (NO) = +\ 90\ kJ\ mol^{-1}$$
$$\Delta H_f^\ominus (NO_2) = +\ 34\ kJ\ mol^{-1}$$

The Hess's Law cycle is

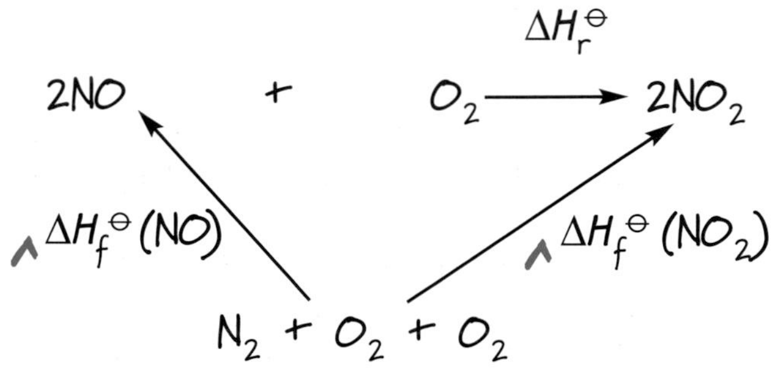

$\Delta H_r^\ominus = \Delta H_f^\ominus (NO_2) - \Delta H_f^\ominus (NO) = +34 - (+90) = -56\ kJ\ mol^{-1}$ ✓ (ecf)

(iii) Construct a reaction profile to represent the reaction in **(ii)**.

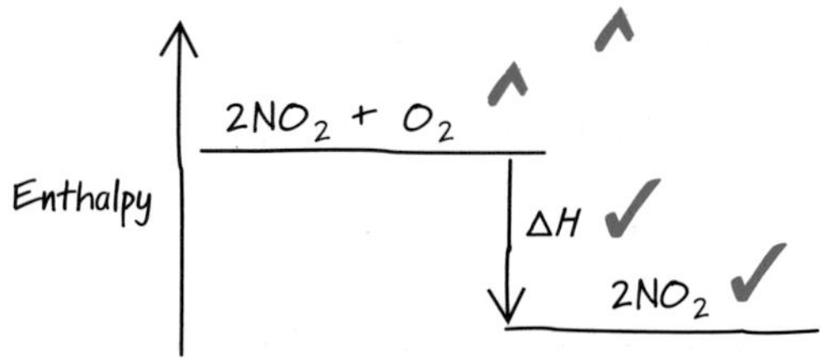

(iv) On the basis of your calculation in **(ii)**, state why you would expect the reaction to take place.

As the reaction is exothermic, the product is thermodynamically more stable than the reactants, and so the reaction should take place. ✓

[9 marks] 6/9

(b) Ethane reacts with chlorine in a substitution reaction:

$$CH_3\text{—}CH_3(g) + Cl_2(g) \rightarrow CH_3\text{—}CH_2Cl + HCl \qquad \Delta H_r^{\ominus} = -115\ \text{kJ mol}^{-1}$$

Given the bond enthalpies below, calculate the bond enthalpy of the C—Cl bond.

Bond	C—H	H—Cl	Cl—Cl
Average bond enthalpy / kJ mol^{-1}	412	431	242

ΔH_r = enthalpy change in bond making – enthalpy change in bond breaking ✗

Bonds broken 1 x C—H and 1 Cl—Cl

Bonds made 1 x C—Cl and 1 x H—Cl ✓

$\Delta H_{break} = 412 + 242 = 654$ $\quad\Delta H_{make} = y + 431$ ✗

$-115 = y + 431 - 654$

$y = -115 - 431 + 654 = +108\ \text{kJ mol}^{-1}$ ✗

7/13 [4 marks] 1/4

2 The standard enthalpy changes of formation of hydrocarbons are difficult to measure directly by experiment, but they can be calculated from standard enthalpy changes of combustion by using Hess's Law.
The table below lists some standard enthalpy changes of combustion of some relevant substances.

Substance	$\Delta H_c^{\ominus}$ / kJ mol^{-1}
$C_3H_8(g)$	–2220
C(s)	–394
$H_2(g)$	–286

(a) (i) Define the term **standard enthalpy of combustion**.

The standard enthalpy of combustion is the enthalpy change when 1 mol ✓ of a substance is ^ burned ^ at 1 atm pressure and at a temperature of 298K. ✓

(ii) Write a balanced equation, including state symbols, to represent the standard enthalpy change of combustion of propane, C_3H_8.

$C_3H_8(g) + 5O_2(g) \rightarrow 3CO_2(g) + 4H_2O(g)$

equation ✓ state symbols ✗

[5 marks] 3/5

(b) The equation that represents the standard enthalpy change of formation, $\Delta H_f^{\ominus}$, of propane is shown below.

$$3C(s) + 4H_2(g) \rightarrow C_3H_8(g)$$

(i) Suggest a reason why $\Delta H_f^{\ominus}$ propane is difficult to determine directly.

Because carbon does not react with hydrogen. ✓

(ii) Use Hess's Law and the data in the table above to calculate a value of $\Delta H_f^{\ominus}$ of propane.

By Hess's Law, $\Delta H_f^{\ominus}(C_3H_8) = \Delta H$ products $- \Delta H$ reactants ✗

$\Delta H_f^{\ominus}(C_3H_8) = -2220 - \{3 \times (-394) + 4 \times (-286)\} = +106$ kJ mol^{-1} ✗

[4 marks] 2/4

(c) In a laboratory experiment, 1.45 g of propanone, C_3H_6O, (molar mass 58 g mol^{-1}) were burned completely in oxygen. The heat from this combustion was used to raise the temperature of 100 g of water from 22.1 °C to 80.2 °C.

(i) Calculate the heat produced. (The specific heat capacity of water is 4.18 J °C^{-1} g^{-1}.)

$\Delta T = 80.2 - 22.1 = 58.1$ °C

Heat q = mass x specific heat capacity x ΔT

$= 1.45 \times 4.18 \times 58.1 = 352.1$ kJ ✗

(ii) Calculate the number of moles of propanone in the 1.45 g.

Moles = mass / molar mass = 1.45g / 58 g mol^{-1} = 0.025 mol ✓

(iii) Use your answers from **(i)** and **(ii)** to calculate the enthalpy of combustion of propanone.

$\Delta H_c(C_3H_6O)$ = heat / moles ✓ = 352.1 / 0.025 = 14100 kJ mol^{-1}

(iv) The value obtained in this way was found to be less than the data book value. Suggest **one** reason why the experimental value is less than the data book value.

Heat was lost to the surroundings ✓ in the experiment, and so the temperature rise was too low.

[6 marks] 3/6

[Total 15 marks] 8/15

How to score full marks

1 (a) (i) All enthalpy definitions refer to **one mole of substance**.

(ii) The student forgot to multiply the $\Delta H_f^{\ominus}$ of both NO and NO_2 by two. This must be done, as there are two moles of each in the equation. Note that $\Delta H_f^{\ominus}$ of all elements in their standard states are **zero**.

(iii) The student drew an energy level diagram. The reaction profile should also include **the reaction pathway** showing the 'hump' due to the activation enthalpy which must also be labelled.

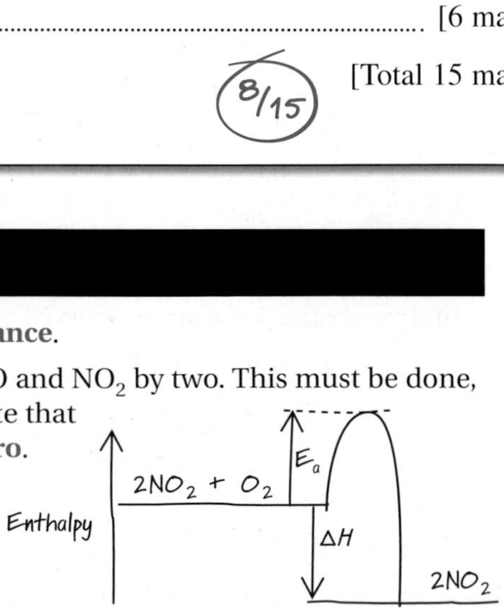

(b) The student made a common error here. Bond **breaking is endothermic** and bond **making is exothermic**. The correct answer is:

ΔH_r = enthalpy change in bond breaking + enthalpy change in bond making

Bonds broken	Bonds made
$1 \times$ C—H $= +412$ kJ mol^{-1}	$1 \times$ C—Cl $= -$ y kJ mol^{-1}
$1 \times$ Cl—Cl $= +242$ kJ mol^{-1}	$1 \times$ H—Cl $= -431$ kJ mol^{-1}

$$-115 = (+412 + 242) + (-y - 431)$$
$$y = +115 + 412 + 242 - 431$$
$$= +338 \text{ kJ mol}^{-1}$$

2 (a) (i) The combustion must be in oxygen and complete, so it should be '...1 mol is **completely** burned in **excess oxygen** ...'

(ii) At 1 atm and 298K, water is a liquid. The reactants and products must be in their standard states in the thermochemical equation.

(b) (ii) Never use expressions such as $\Delta H = \Delta H$ products $- \Delta H$ reactants. It works to calculate $\Delta H_{reaction}$ when $\Delta H_{formation}$ data is supplied, but it will never score a mark. **Draw a Hess's Law cycle and then use it to do the calculation.** The student was awarded one mark for multiplying $\Delta H_c^{\ominus}$ of C and H_2 by 3 and 4 respectively. The correct cycle and calculation are:

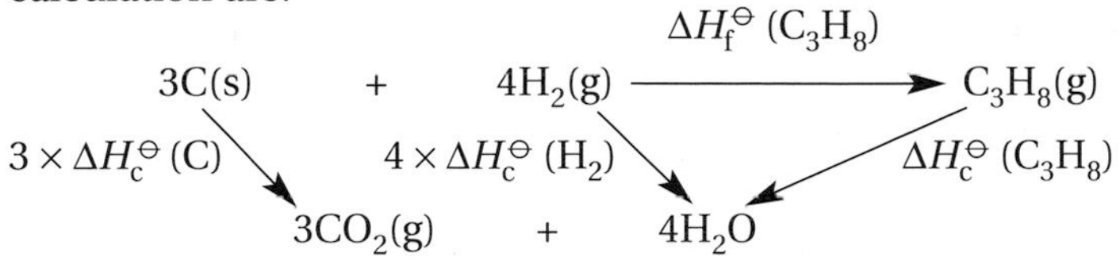

Cycle ✓

$\Delta H_f^{\ominus}(C_3H_8) + \Delta H_c^{\ominus}(C_3H_8) = 3 \times \Delta H_c^{\ominus}(C) + 4 \times \Delta H_c^{\ominus}(H_2)$ ✓

$\Delta H_f^{\ominus}(C_3H_8) = 3 \times (-394) + 4 \times (-286) - (-2220) = -106$ kJ mol^{-1} ✓

(c) (i) The mass in the expression is **the mass of the water that heated up**, not the mass of the reactant. The other error was in the **units**. As the specific heat capacity is measured in joules, the unit of heat will also be in joules. The answer should be:

heat, q = 100 g $\times$ 4.18 J ^{o}C^{-1} g^{-1} $\times$ 58.1 ^{o}C = 24300J = 24.3 kJ

(iii) As the temperature rose, the reaction is **exothermic** and so the value of ΔH will be **negative**. The correct answer is $\Delta H_c^{\ominus} = -24.3$ kJ / 0.025 mol $= -972$ kJ mol^{-1}.

Don't make these mistakes ...

Don't forget that all definitions refer to 1 **mole of substance.**

Don't forget to include **state symbols** in all thermochemical equations.

The 'formula' $\Delta H_r = \Sigma\Delta H$ of products $- \Sigma\Delta H$ of reactants is only true when using enthalpy of formation data, and, if used, must be written as $\Delta H_r = \Sigma\Delta H_f$ of products $- \Sigma\Delta H_f$ of reactants. **It is always better to draw and use a Hess's Law cycle.**

Don't forget to learn the **definitions.**

In questions based on experimental data from a calorimeter, remember that the mass in the expression 'mass $\times$ specific heat capacity $\times \Delta T$ is the **total** mass of the substances that heat up (or cool down). This includes the solvent.

Don't forget to **balance** a Hess's Law cycle.

Don't get the sign wrong. ΔH = minus the heat produced per mole. If the temperature rises, the reaction is exothermic and so ΔH is negative. If the temperature falls, the reaction is endothermic and ΔH is positive.

Key points to remember

Standard conditions must be mentioned in definitions. They are:

- A constant pressure of 101 kPa (1 atm)
- A stated temperature (usually 25 °C)
- All solutions at a concentration of 1 mol dm^{-3}

Exothermic reactions: the temperature rises, so heat is given out to the surroundings.

ΔH is negative.

Endothermic reactions: the temperature falls, so heat is then taken in from the surroundings.

ΔH is positive

Standard enthalpy of formation is the enthalpy change when **1 mole** of a compound is formed from its **elements** under **standard conditions**. The standard enthalpy of formation of copper sulphate is the enthalpy change for the reaction:

$$Cu(s) + S(s) + 2O_2(g) \rightarrow CuSO_4(s)$$

Standard enthalpy of combustion is the enthalpy change when **1 mole** of a substance is burnt in **excess oxygen** under **standard conditions**. All ΔH_c values are **exothermic**. The standard enthalpy of combustion of ethanol, C_2H_5OH, is the enthalpy change for the reaction:

$$C_2H_5OH(l) + 3O_2(g) \rightarrow 2CO_2(g) + 3H_2O(l)$$

Standard enthalpy of neutralisation is the enthalpy change when an acid reacts with a base in 1 mol dm^{-3} solutions to **form 1 mole of water**. For a strong acid being neutralised by a strong base, it is the enthalpy change for the reaction:

$$H^+(aq) + OH^-(aq) \rightarrow H_2O(l)$$

Hess's Law states that the enthalpy change for a reaction is independent of the route by which the reaction is achieved.

In the cycle below:

$\Delta H = \Delta H_1 + \Delta H_2 + \Delta H_3$

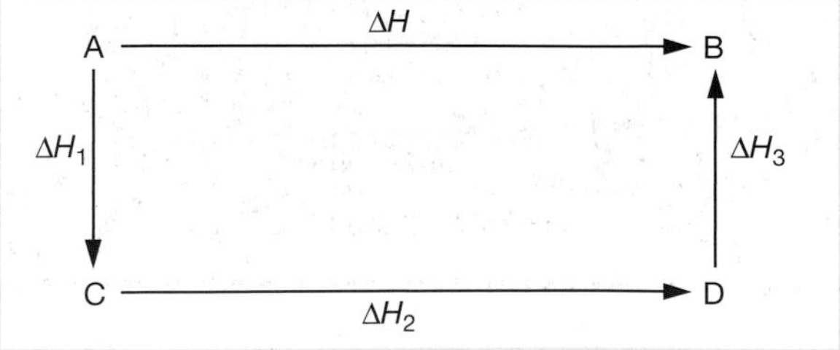

One common cycle is using **enthalpies of formation**:

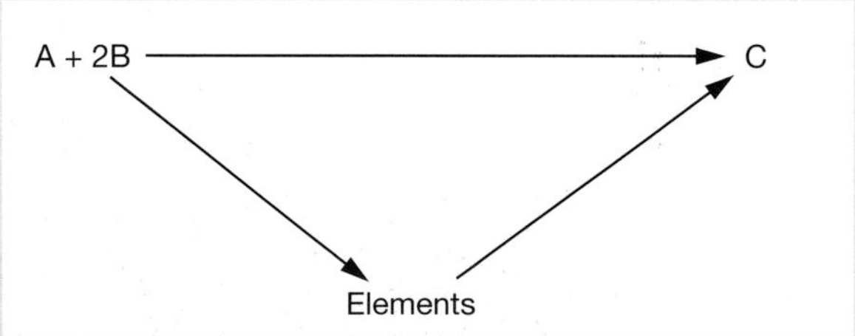

$\Delta H_r = -\Delta H_f$ of A $- 2x\ \Delta H_f$ of B $+ \Delta H_f$ of C

Another common cycle is using **enthalpies of combustion**:

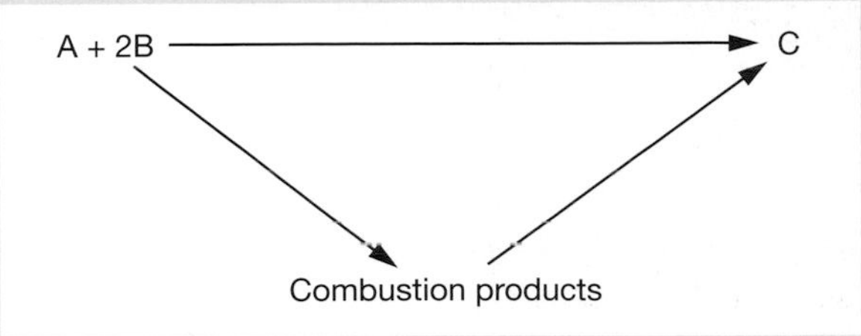

$\Delta H_r = \Delta H_c$ of A + 2x ΔH_c of B – ΔH_c of C

Bond enthalpy is the enthalpy change per mole of bond broken. In calculations you need to use structural formulae to see what bonds are broken and what bonds are made.

- Bond breaking is endothermic (+)
- Bond making is exothermic (–)

ΔH reaction = enthalpy of bonds broken (a positive number) + enthalpy of bonds made (a negative number).

Calorimetry

Heat produced, q = m. c. ΔT

heat produced per mole = q/moles reacting

For **exo**thermic reactions ΔH is **negative**.

For **endo**thermic reactions ΔH is **positive**.

m = **total** mass of the contents of the calorimeter (solutions are assumed to have a mass equal to their volume).

c = the specific heat capacity (its value will be given in the question)

ΔT is the change in temperature.

Enthalpy level diagrams show the relative energy levels of the reactants (on the left) and of the products (on the right) with ΔH shown by an arrow. For the reaction

$A + B \rightarrow C \quad \Delta H = -123 \text{ kJ mol}^{-1}$

The diagram is shown below:

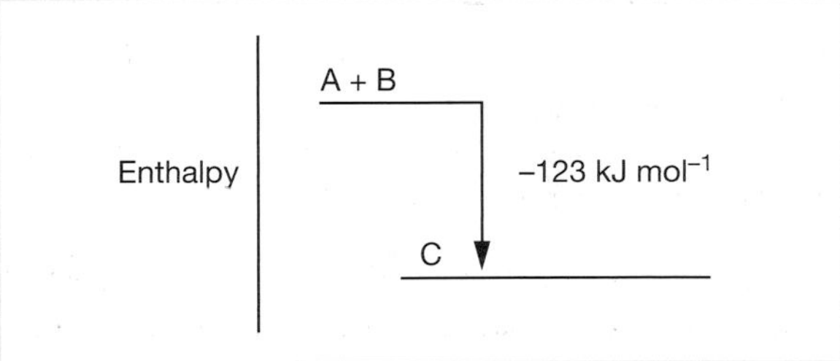

BOND BREAKING IS ENDOTHERMIC

BOND MAKING IS EXOTHERMIC

Questions to try

Examiner's hints ***for question 1***

(a) (ii) It is sensible to add an equation with state symbols in all thermochemical equations.

(b) The alternative route in the cycle is via the elements.

(c) Draw structural formulae to see exactly what bonds are being broken and what bonds are being made.

Q1

(a) Define:

(i) standard conditions

....

....

[3 marks]

(ii) the standard enthalpy of formation of a substance such as ethene, $CH_2{=}CH_2$.

....

....

[3 marks]

(b) Draw a Hess's Law cycle. Use it and the standard enthalpies of formation given in the table below to calculate the standard enthalpy of the reaction:

$$CH_2{=}CH_2(g) + H_2O(g) \rightarrow CH_3{-}CH_2OH(g)$$

	$CH_2{=}CH_2(g)$	$H_2O(g)$	$CH_3{-}CH_2OH(g)$
$\Delta H_f^\ominus$ / kJ mol^{-1}	+ 53	– 242	– 235

[3 marks]

(c) (i) Use the average bond enthalpies given below to calculate the standard enthalpy change for the same reaction.

	C—H	C═C	H—O	C—C	C—O
Bond enthalpy / kJ mol^{-1}	412	612	463	348	360

[3 marks]

(ii) Explain why the answer to (c) (i) is different from that obtained in (b).

....

[2 marks]

[Total 14 marks]

Examiner's hints ***for question 2***

(a) When using your graph to evaluate ΔT, remember that the reaction started at 3 minutes.

(b) What is the total mass that warmed up? Is the reaction exo- or endothermic?

(c) The aq in the equations stands for many moles of water, and so includes the $5H_2O$ that is in the final equation.

Q2

Water of mass 100 g was placed in a polystyrene cup and its temperature measured. After three minutes 4.80 g of anhydrous copper sulphate, $CuSO_4(s)$, was added and the mixture was thoroughly stirred for two minutes, by which time all the solid had dissolved. The temperature was measured for a further five minutes. The following results were obtained:

Temperature / °C	15.2	15.2	15.2	20.1	19.9	19.7	19.5	19.3	19.2
Time / minutes	0	1	2	5	6	7	8	9	10

Plot a graph of temperature (y-axis) against time (x-axis) and use it to determine the temperature change, ΔT.

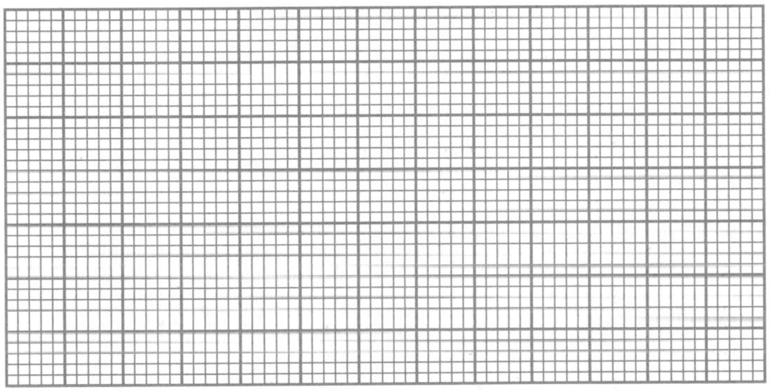

[5 marks]

(b) Hence calculate the molar enthalpy of solution of anhydrous copper sulphate

$$CuSO_4(s) + aq \rightarrow CuSO_4(aq)$$

given that the specific heat capacity of the solution is 4.18 J g^{-1} °C^{-1} [4 marks]

(c) The molar enthalpy of solution of hydrated copper sulphate

$$CuSO_4.5H_2O(s) + aq \rightarrow CuSO_4(aq)$$

is + 4.7 kJ mol^{-1}. Draw a Hess's Law cycle or an energy level (profile) diagram and use it to calculate the enthalpy change for the reaction:

$$CuSO_4(s) + 5H_2O(l) \rightarrow CuSO_4.5H_2O(s)$$ [3 marks]

[Total 12 marks]

The answers to these questions are on pages 85 and 86.

Exam Questions and Student's Answers

1 (a) Describe the collision theory of rates of reaction.

For a reaction to take place, the molecules or ions must collide. ✓ The particles are travelling at different speeds and they must collide hard enough to break the bonds. ✗ The more frequent the collisions, the faster is the rate of the reaction. ✓ The larger the fraction that collide hard enough and so react on collision, the faster is the reaction. ✓ (ecf)

[5 marks] 3/5

(b) The rate of the reaction between pieces of solid calcium carbonate and dilute hydrochloric acid was studied by measuring the time taken to produce 10 cm^3 of carbon dioxide gas. The equation for the reaction is:

$$CaCO_3(s) + 2HCl(aq) \rightarrow CaCl_2(aq) + H_2O(l) + CO_2(g)$$

Experiments were carried out using identical pieces of calcium carbonate, but acid of concentration either 1.0 mol dm^{-3} or 0.10 mol dm^{-3}, and temperatures of either 15 °C or 25 °C. The results obtained are shown in the table below.

Experiment	Time / s	Concentration / mol dm^{-3}	Temperature / °C	
1	8.0	0.10	25	✗
2	39	1.0	15	✗
3	4.5	1.0	25	✓
4	82	0.10	15	✓

(i) State which was the fastest experiment.

Experiment number 3. ✓

(ii) Insert the values for the concentrations and temperatures in the table above.

(iii) State and explain what effect, if any, would an increase in pressure have on the rate of this reaction?

An increase in pressure would cause the reaction to slow down, ✗ because there is gas on the right hand side of the equation. ✗

[6 marks] 3/6

6/11 [Total 11 marks]

2 The distribution of molecular energies at a temperature T_1 is shown below.

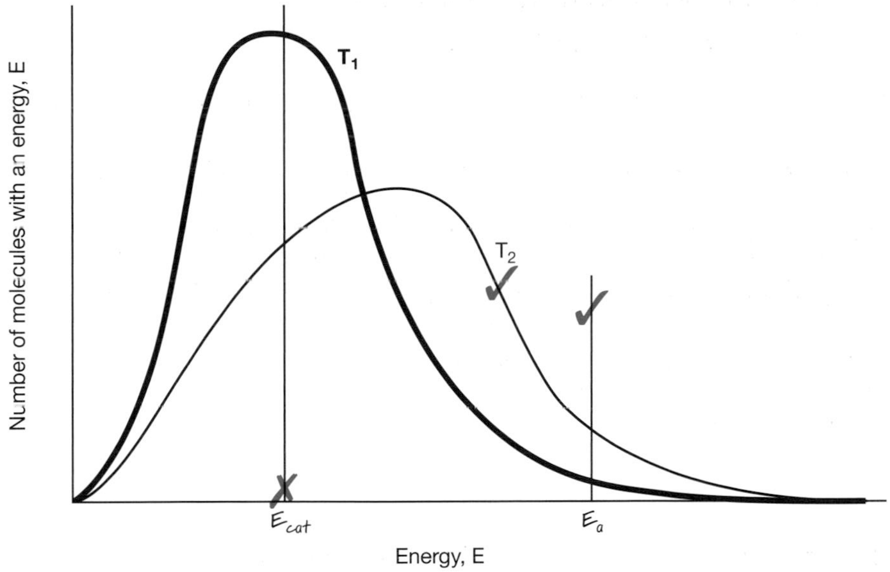

(a) Sketch a graph, labelling it T_2, showing the distribution of molecular energies at a higher temperature T_2 and hence explain why the rate of reaction is greater at temperature T_2 than at T_1.

At T_2 there are more molecules that have energy greater ✓ than the activation energy, E_a. ^ ^

[5 marks] 3/5

(b) Explain the effect of a catalyst on the rate of a reaction. You should use the graph in your answer.

^ The activation energy for the catalysed route is lower ✓ than that for the uncatalysed reaction. ^

[4 marks] 1/4

4/9 [Total 9 marks]

How to score full marks

1 (a) The two important points are the frequency of collision and the energy that the colliding particles have. The particles have different **energies** and only a small fraction has **energy ≥ activation enthalpy**, and so are able to react when they collide. It is too vague to say that they must collide hard enough and wrong to say that the bonds have to be broken. New bonds are being made (exothermic) as the old bonds break (endothermic). The student completed the answer well, by relating the frequency of collision and the fraction of collisions that result in reaction to the rate of the reaction.

(b) **(ii)** The student correctly identified that the fastest rate would be in the experiment with the highest acid concentration and temperature, and that the slowest rate would be when the concentration and the temperature had the lowest value. However, a 10 × increase in the concentration of a reactant will cause **a greater increase in rate** than a 10 °C rise in temperature, so experiment 1 had a concentration of 1.0 mol dm^{-3} and a temperature of 15 °C.

(iii) Pressure only affects the rate where a **reactant** is a gas. In this reaction CO_2 is a product, and so a change of pressure will have **no** effect on the rate of the reaction.

2 **(a)** Marks were lost because:

- the student should have stated **why** the graph shows that more molecules have energy greater than the activation energy. The correct answer is **because the area under the graph to the right of the E_a line is greater for the T_2 graph than for the T_1 graph.**
- the student should also have said that this means that **more of the collisions will result in reaction** (or more of the collisions will be successful).

The student scored the first mark, which is for the shape of the graph for T_2. The peak for T_2 **must be to the right and lower than the peak for T_1** and **the area under the two curves should be the same**, but if you draw it badly write 'The areas under the two graphs should be the same', **then you will not be penalised.**

The student gained the second mark because the activation energy for the uncatalysed reaction was correctly drawn to the right of the peaks at both temperatures.

The third mark was gained for saying that more molecules have energy greater than the activation energy.

(b) The student lost a mark because the line for the catalysed activation energy, E_{cat}, was drawn in the wrong place on the graph. **The line must be to the right of both peaks.**

The student failed to state that **the rate of reaction would be faster with a catalyst.**

The student should also have explained that the **lower activation energy** means that a **greater proportion of the molecules** will have the necessary energy to **react on collision.**

Don't make these mistakes ...

Don't state that increasing the temperature brings about an increase in the rate of a reaction by causing a greater collision frequency. A temperature increase of 10 °C causes only a very slight increase in the frequency of collision (about a 1% increase). **The important point is that this temperature rise causes a 100% increase in the fraction of molecules that have the activation energy to react on collision.**

Don't say that a catalyst speeds up a reaction by lowering the activation energy. Homogeneous catalysts (those that are in the same state as the reactants, e.g. Fe^{2+} ions in the oxidation of iodide ions by persulphate ions) work by combining with one of the reactants to form an intermediate, which then reacts to reform the catalyst and the products. **This alternate reaction route has a lower activation energy.**

Heterogeneous catalysts (those that are in a different state to the reactants, e.g. iron in the Haber process for the manufacture of ammonia from nitrogen and hydrogen gases) adsorb the reactants on their surface. The adsorbed gases react to form the products and then leave the surface of the catalyst. **This alternate reaction route also has a lower activation energy.**

Key points to remember

Maxwell-Boltzmann distribution

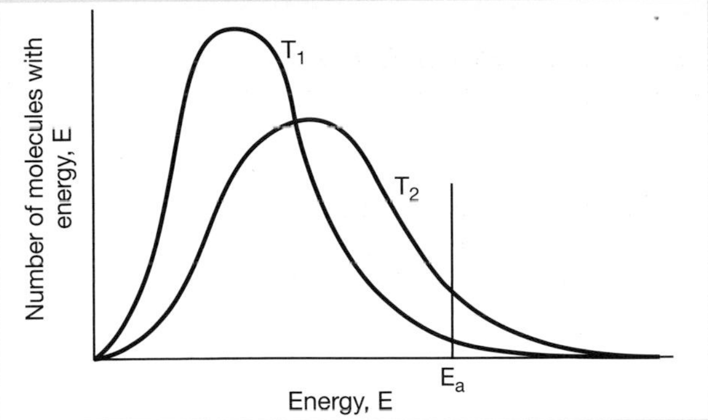

This shows the way in which the kinetic energy of the molecules in a gas (or liquid) are distributed at a temperature T_1. The number of molecules that have energies greater than or equal to a value E_a is equal to the **area under the curve** to the right of the E_a line.

At a higher temperature T_2, a larger number have energies greater than or equal to this value.

Factors that affect the rate of reaction

Factor	How it affects the rate	Why it affects the reaction
Temperature	An increase in temperature increases the rate of a chemical reaction. This is true for both exothermic and endothermic reactions. 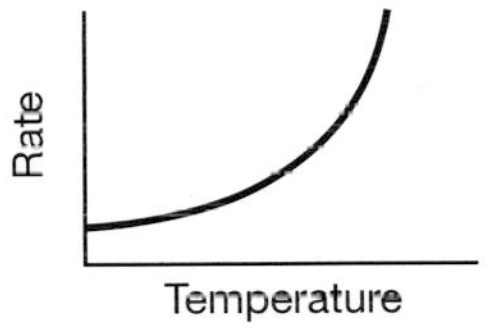	The particles have more kinetic energy so more of them have energy ≥ the activation energy. Thus more of them react on collision. The collision frequency also increases but slightly (see diagram above).
Concentration	An increase in the concentration of substances in solution increases the rate of reaction.	The particles are closer together and so the frequency of collision is increased.
Pressure	An increase in pressure of a gas increases the rate of reaction.	As above, the gas molecules are closer together and so the frequency of collision is increased.
Catalyst	A catalyst speeds up a reaction without getting used up. 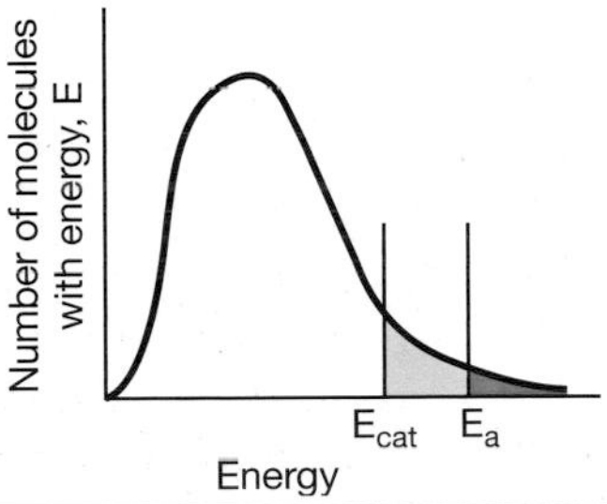	A catalyst provides an alternative reaction route which has a lower activation energy, thus , at a given temperature, more particles will have kinetic energy greater than or equal to the activation energy. 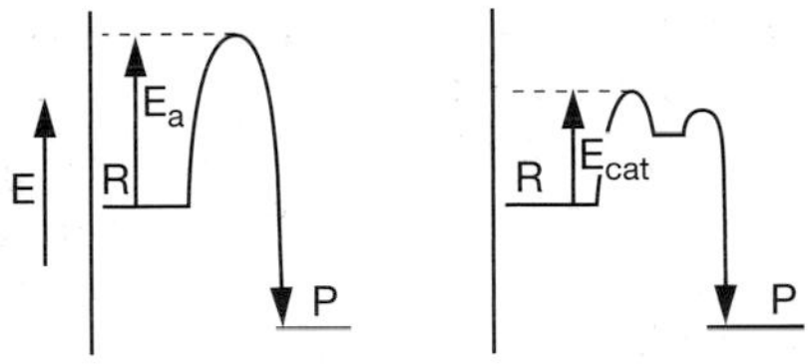
Surface area	The rate of a reaction of a solid with a gas or a liquid is dependent on the surface area of the solid – the bigger the surface area the faster the reaction.	The frequency of collision of a reactant molecule or ion with the surface is increased.

Questions to try

Examiner's hint ***for question 1***
There are 5 factors which may affect the rate of a reaction. Which are applicable in each reaction?

Q1

(a) An aqueous solution of 1-chloropropane reacts slowly with dilute sodium hydroxide at 25 °C according to the equation:

$$CH_3CH_2CH_2Cl\ (aq) + OH^-(aq) \rightarrow CH_3CH_2CH_2OH(aq) + Cl^-(aq)$$

State three ways by which the rate of this reaction could be increased.

....

....

....

[3 marks]

(b) Explain why the flour dust in a flour milling factory may explode, whereas it is difficult to ignite ordinary flour in the home.

....

....

....

[3 marks]

(c) Consider the reaction: $Zn(s) + 2HCl(aq) \rightarrow ZnCl_2(aq) + H_2(g)$

Explain the effect, if any, on the rate of reaction of:

(i) increasing the pressure

....

[2 marks]

(ii) adding water

....

[2 marks]

(iii) Using hydrobromic acid (HBr), of the same concentration, instead of hydrochloric acid.

....

....

[2 marks]

[Total 12 marks]

Examiner's hint *for question 2*

For questions involving the Maxwell-Boltzmann distribution, you may need to draw two graphs at different temperatures. These must be clearly labelled. Draw in a suitable activation energy and use collision theory to **state** and **explain** the effect on the rate.

Q2

(a) Define activation energy.

[2 marks]

(b) Nitrogen(V) oxide decomposes slowly at room temperature according to the equation:

$$2N_2O_5(g) \rightarrow 4NO_2(g) + O_2(g)$$

(i) Draw Maxwell-Boltzmann distribution of energies graphs at two different temperatures, and use them to explain how an increase of temperature affects the rate of the reaction.

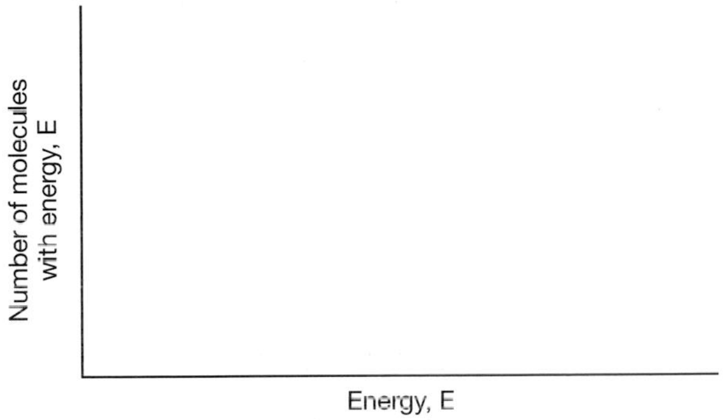

[8 marks]

(ii) Use collision theory to explain how the rate of this reaction would be altered by an increase in pressure.

[3 marks]

[Total 13 marks]

The answers to these questions are on pages 86 and 87.

Exam Question and Student's Answer

1 (a) What is meant by the term "dynamic equilibrium"?

The system reacts rapidly until the substances reach equilibrium, when there is no further reaction. ✗

[2 marks] 0/2

(b) In the Haber process hydrogen is manufactured from methane, air and water. The first step in this process is the reaction between methane, CH_4, and steam at a temperature of 1000 K in the presence of a nickel catalyst at a pressure of 30 atmospheres.

$$CH_4(g) + H_2O(g) \rightleftharpoons CO(g) + 3H_2(g) \qquad \Delta H = +206\,\text{kJ mol}^{-1}$$

State and explain the effect on the position of equilibrium of the pressure being reduced to a lower value at the same temperature.

A decrease in pressure will drive the reaction ^ to the side with more ^ molecules.

[2 marks] 0/2

(c) In the second step the carbon monoxide is oxidised by water at 700 K in the presence of an iron oxide catalyst.

$$CO(g) + H_2O(g) \rightleftharpoons CO_2(g) + H_2(g) \qquad \Delta H = -41\,\text{kJ mol}^{-1}$$

(i) State and explain the effect on the position of equilibrium of increasing the temperature at constant pressure.

An increase in temperature will drive the equilibrium to the left, ✓ as the reaction is exothermic ^.

[2 marks] 1/2

(ii) Why is a temperature of 700 K used rather than a higher or a lower value?

If the temperature were higher the yield would be less, ✓ but if the temperature were lower, although the yield would be greater, ✓ the rate of reaction would be much less ✓ ^.

[4 marks] 3/4

(iii) Explain why a catalyst is used in this manufacturing process.

A catalyst increases the rate of reaching equilibrium, and so allows the reaction to take place quickly at a temperature, which is not too high and expensive, and at which there is a reasonable yield.

[3 marks] 2/3

(d) Hydrated cobalt(II) ions, which are pink, react reversibly with a solution of chloride ions:

$$\underset{\text{pink}}{[Co(H_2O)_6]^{2+}} + 4Cl^- \rightleftharpoons \underset{\text{blue}}{[CoCl_4]^{2-}} + 6H_2O$$

When a solution containing blue $[CoCl_4]^{2-}$ ions is cooled in iced water, the solution turns pink. Deduce whether the reaction as written above is exo or endothermic.

A decrease in temperature drives the reaction to the left, and so this must be the endothermic direction. Therefore the reaction as written is endothermic left to right.

9/16 [3 marks] 3/3

[Total 16 marks]

How to score full marks

1 (a) The student's answer is wrong. **When a question says 'what is meant by...' you must explain both words – dynamic and equilibrium.** The correct answer is:

'Dynamic equilibrium is when the rate of the forward reaction equals the rate of the reverse reaction, and so there is no further change in the concentrations of any of the substances in the equilibrium mixture.'

(b) The student is asked to state and explain the effect of a pressure decrease but fails to do this. **The effect is that the position of equilibrium is driven to the right, and the explanation is that the right-hand side contains more gas molecules than the left.** (This can be predicted by Le Chatelier's principle – see Key Points on page 64.)

(c) (i) The student correctly stated the effect of increasing the temperature, but the explanation was slightly ambiguous, so only 1 mark was given. **An equilibrium reaction goes in both directions, so it is important to make it clear which direction is exothermic and which endothermic.** The correct answer is:

'The position of equilibrium shifts to the left, because an increase of temperature will drive the equilibrium in the endothermic direction, which is from right to left.'

(ii) **There are four marks for this question, but the student only made three points** and so scored 3 marks. The student missed out the important point that the temperature of 700 K is a compromise temperature giving a reasonable yield at a good rate.

(iii) The student omitted the vital fact that **a catalyst has no effect on the position of equilibrium.**

Key points to remember

Dynamic equilibrium

There is no further change in the concentrations of any of the chemicals, because the rate of the forward reaction equals the rate of the reverse reaction.

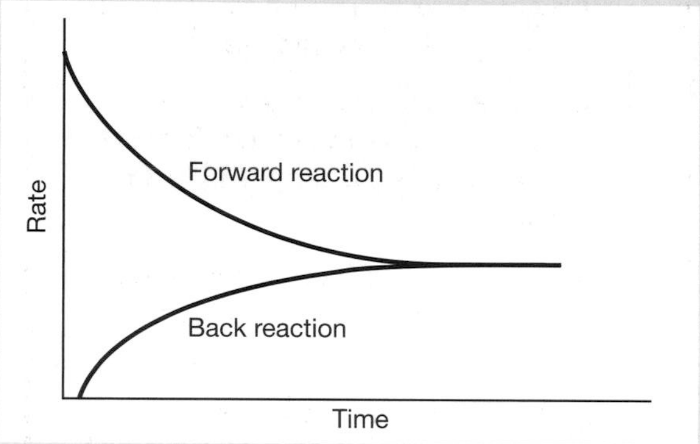

Le Chatelier's principle is: "If the conditions of a system at equilibrium are altered, the system will react as if to restore the original conditions".

This principle can be used to predict changes in position of equilibrium. It does not explain them.

When a reaction is in equilibrium, an **increase in temperature** will:

- move the position of equilibrium in the **endothermic** direction.
- increase the rate of the forward *and* the back reactions. However, the rate of the endothermic reaction is increased more than the rate of the exothermic reaction.
- mean that equilibrium is reached sooner at higher temperatures.

An example:

$N_2(g) + 3H_2(g) \rightleftharpoons 2NH_3(g) \quad \Delta H = -92\,kJ\,mol^{-1}$

- The reaction is exothermic left to right.
- A temperature increase causes the position of equilibrium to move to the left.
- A temperature decrease causes the position of equilibrium to move to the right, but will make the reaction slower.
- A compromise temperature of 400 °C is used in the manufacture of ammonia.

- When a reaction is in equilibrium, an **increase in pressure** will move the position of equilibrium to the side with **fewer gas molecules**.

  An example:
  $N_2(g) + 3H_2(g) \rightleftharpoons 2NH_3(g)$
  4 gas molecules $\rightleftharpoons$ 2 gas molecules
- A high pressure increases the rate (slightly).
- A very high pressure is very expensive.
- An increase in pressure will move the position of equilibrium to the right. As the yield at ordinary pressures is very low, a pressure of 200 atmospheres is used.

The effect of a catalyst

A catalyst has no effect on the position of equilibrium; it speeds up the rate of the forward and the back reactions equally, and so equilibrium is reached more quickly.

- The catalyst for the Haber process for manufacturing ammonia is iron.
- The catalyst for the contact process for manufacturing sulphuric acid is vanadium(V) oxide.

Removal of a reactant

If a reactant is removed physically or chemically, the position of equilibrium will shift to make more of that reactant.

If sodium hydroxide solution is added to the equilibrium:

$$CH_3COOC_2H_5 + H_2O \rightleftharpoons CH_3COOH + C_2H_5OH$$

it removes the ethanoic acid, CH_3COOH, and drives the equilibrium to the right.

Increase in concentration of a reactant

If the concentration of a reactant is increased, the position of equilibrium will shift to remove that reactant.

If concentrated HCl is added to the equilibrium:

$$PbCl_2(s) + 2Cl^-(aq) \rightleftharpoons PbCl_4^{2-}(aq)$$

the concentration of Cl^- ions is increased and so the equilibrium is shifted to the right. Thus the solid $PbCl_2$ will dissolve.

Economic factors

The aim of an industrial process is a high yield made quickly and cheaply.

- A high pressure is very expensive.
- A very high temperature is expensive.
- Unused reactants are recycled through the reaction vessel.
- Exothermic processes are carried out at a temperature that is a compromise between yield and rate.
- A catalyst enables the reaction to proceed at a temperature that gives a good yield quickly.
- A high pressure (other than that required to push the gases through the plant) is only used if the yield is otherwise very low.

Don't make these mistakes ...

Don't think that **all reactions cease** when equilibrium has been reached.

Don't forget that **equilibrium can only be reached in a closed system**, which is one in which no chemicals enter or leave the reaction mixture.

Don't forget to discuss rate as well as yield (or position of equilibrium) when answering a question on the optimum industrial conditions.

Don't muddle pressure with volume. For equilibrium reactions that involve gases, an increase in volume will cause a decrease in pressure, which will then result in the position of equilibrium being moved to the side with more gas molecules.

Question to try

Examiner's hints

1 (c) and **(d)**

Remember that Le Chatelier's principle predicts what happens if the conditions of a system in equilibrium are altered.

High pressure is expensive and is only used for low yield reactions.

Q1

(a) Gases A and B were mixed and allowed to reach equilibrium in a closed container:

$$A(g) + B(g) \rightleftharpoons C(g)$$

Draw on the axes below how the rates of the forward and back reactions alter over time until equilibrium is reached.

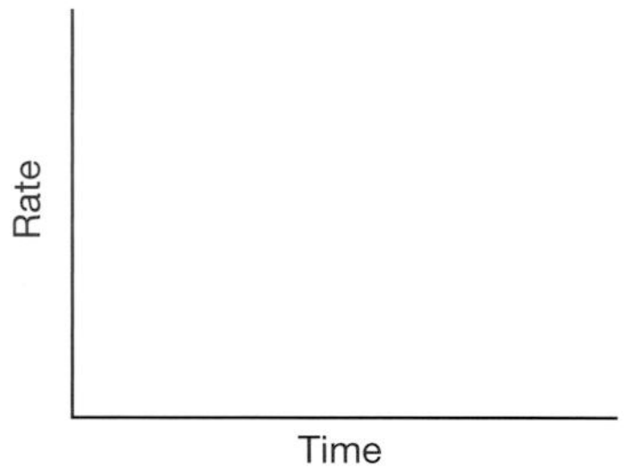

[3 marks]

(b) Which statements about a system in equilibrium are true and which are false?

No reactions are occurring. ....

The concentrations of the substances do not alter. ....

The amount of product equals the amount of reactant. ....

The rate of the forward reaction equals the rate of the reverse (back) reaction. ....

[4 marks]

(c) The critical step in the manufacture of sulphuric acid is the oxidation of sulphur dioxide.

$$SO_2(g) + \tfrac{1}{2}O_2(g) \rightleftharpoons SO_3(g) \qquad \Delta H = -98\,\text{kJ mol}^{-1}$$

Explain why the following conditions are chosen to make the process as economic as possible.

(i) A temperature of 425 °C

....

....

....

[4 marks]

(ii) A pressure of 202 kPa (2 atm)

....

....

[2 marks]

(iii) A catalyst of vanadium(V) oxide

....

....

....

[3 marks]

(d) Yellow chromate(VI) ions in aqueous solution are in equilibrium with orange dichromate(VI) ions.

$$2CrO_4^{2-}(aq) + 2H^+(aq) \rightleftharpoons Cr_2O_7^{2-}(aq) + H_2O(l)$$

State and explain what you would see when:

(i) a concentrated solution of sodium hydroxide is added to a solution containing dichromate(VI) ions.

....

....

[3 marks]

(ii) dilute sulphuric acid is added to a solution of chromate(VI) ions.

....

....

[3 marks]

[Total 22 marks]

The answers to this question are on page 88.

Exam Questions and Student's Answers

1 Consider the reaction scheme set out below:

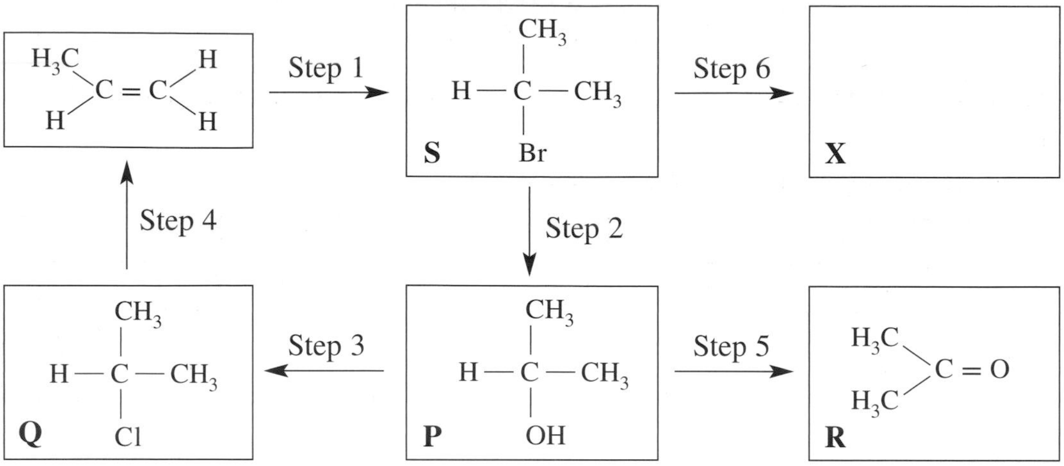

(a) From compounds **S**, **P**, **Q** and **R** identify by letter, which, if any, is:

(i) a ketone

Substance R is a ketone ✓

(ii) a member of the same homologous series as butan-1-ol.

None ✗ as butan-1-ol is a primary alcohol and substance P is a secondary alcohol.

[2 marks] 1/2

(b) (i) Identify the reagent necessary for Step 3.

Phosphorus pentachloride ✓

(ii) Steps 2 and 4 both have the same reagent which is potassium hydroxide. In what way do the conditions vary between the two steps?

Step 4 requires the ethanolic ✓ solution to be heated and Step 2 happens at room temperature. ✗

(iii) Identify the reagents and conditions for Step 5.

Reagents: acidified dichromate ✗

Conditions: heat under reflux ✓

(iv) Two possible isomers could be produced in Step 1. One is shown in the scheme. Identify the other isomer and explain why 2-bromopropane is the major product.

Other isomer: $CH_3CH_2CH_2Br$ ✓

Reason: Because Markovnokov's rule states that the hydrogen goes onto the carbon atom which already has more hydrogen atoms directly attached to it. ✗

[7 marks] 4/7

(c) Propene can be polymerised to poly(propene). Draw a section of this polymer showing **two** repeat units.

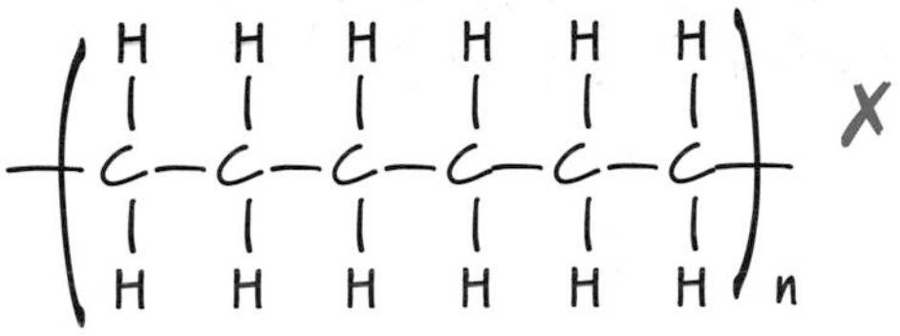

[1 mark] 0/1

(d) Compound **X** has the composition by mass of 69.6% carbon, 10.1% hydrogen and 20.3% nitrogen.

(i) Calculate the empirical formula of compound **X**.

	%	% ÷ A_r	÷ 1.45
Carbon	69.6	69.6 ÷ 12 = 5.8	5.8 ÷ 1.45 = 4
Hydrogen	10.1	10.1 ÷ 1 = 10.1	10.1 ÷ 1.45 = 6.97
Nitrogen	20.3	20.3 ÷ 14 = 1.45 ✓	1.45 ÷ 1.45 = 1 ✓

The empirical formula is C_4H_7N. ✓

(ii) Suggest the reagent necessary to convert compound **S** to compound **X** in Step 6.

Ammonia X

[4 marks] 3/4

8/14 [Total 14 marks]

2 Chlorine reacts with both ethane and ethene.

(a) Write the equation for the reaction of ethane with chlorine, and classify the type of reaction occurring.

Equation: $C_2H_6 + Cl_2 \rightarrow C_2H_5Cl + HCl$ ✓

Type of reaction: ^ Substitution [2 marks] 1/2

(b) State the necessary condition for this reaction.

Ultraviolet light ✓ [1 mark] 1/1

(c) Write the equation for the reaction of ethene with chlorine, and classify the type of reaction occurring.

Equation: $C_2H_4 + Cl_2 \rightarrow C_2H_4Cl_2$ X

Type of reaction: Electrophilic addition ✓ [2 marks] 1/2

3/5 [Total 5 marks]

3 (not for Edexcel candidates)

(a) (i) Outline, using curly arrows, the mechanism for the reaction between aqueous hydroxide ions and iodoethane.

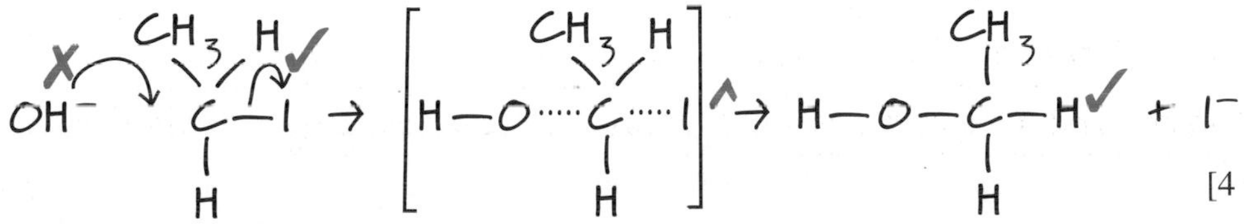

[4 marks] 2/4

(ii) Classify this mechanism.

It is a nucleophilic ✓ substitution ✓ reaction. [2 marks] 2/2

(b) Ammonia reacts with iodoethane by a similar mechanism. What, in the ammonia molecule, makes this possible?

The nitrogen atom is very δ−. ✗ [1 mark] 0/1

4/7 [Total 7 marks]

How to score full marks

1 (a) (ii) Primary and secondary alcohols are both members of the homologous series of alcohols. Members of the same series must have **the same general formula** (here both are $C_nH_{2n+1}OH$) and have **the same functional group** (here both have an OH group).

(b) (ii) The elimination reaction in Step 4 requires the KOH to be in an **ethanolic** solution, and the substitution reaction in Step 2 requires an **aqueous** solution. Both need to be heated under reflux.

(iii) The full name or formula of reagents must be given. So the reagent is potassium dichromate(VI) and dilute sulphuric acid.

(iv) Markovnikov's rule does not **explain** the direction of addition, it only **predicts** it. The correct answer is:

If the hydrogen adds on to the end carbon atom, the intermediate **secondary carbocation** formed is **more stable** than the **primary** carbocation which would be the intermediate obtained if the hydrogen added onto the middle carbon atom.

(c) The answer is wrong. It should be:

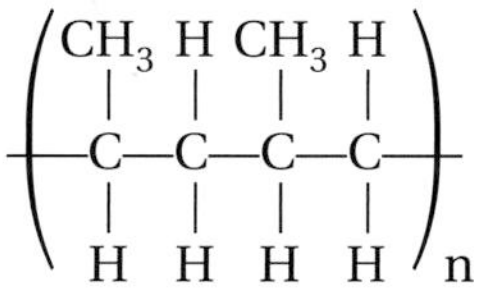

Atoms are not rearranged during polymerisation, and so the CH_3 groups will be on the side chain.

(d) (ii) Nitrogen atoms can be put into organic molecules either with ammonia or with cyanides. Here the number of carbon atoms has **increased** from three in substance **S** to four in substance **X**. The reagent is **potassium cyanide**, KCN.

2 (a) The correct answer is **free radical substitution**.

(c) The formula $C_2H_4Cl_2$ is ambiguous. It could be CH_3CHCl_2, which is wrong here, or **CH_2ClCH_2Cl**, which is correct.

3 (a) (i) There are two errors in the mechanism. The curly arrow **must start from the lone pair of the <u>oxygen</u> atom in the OH^- ion**. The second error is that the student **omitted the negative charge on the transition state** (intermediate). The correct mechanism is:

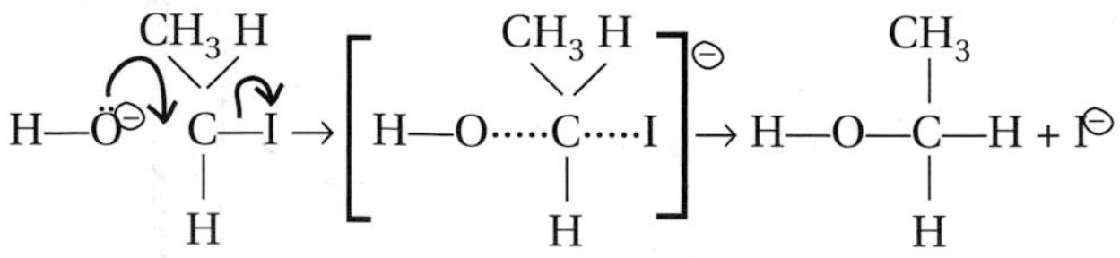

(b) Nucleophiles must have a lone pair of electrons which they use to form a covalent bond (in this case with the carbon atom). The correct answer is:

'The nitrogen atom in ammonia has a **lone pair of electrons** as does the oxygen in the hydroxide ion.'

Don't make these mistakes ...

Don't forget to define nucleophiles and electrophiles in terms of donating or accepting a lone pair of electrons and hence forming a covalent bond.

When describing the test for alkenes, don't forget to give the **colour** of the bromine solution **before** and **after** the reaction.

In mechanisms, ensure that a **curly arrow** starts from a bond or a lone pair, and ends forming an anion or a bond.

Always write cis/trans isomers with the correct **120°** bond angle around the double bonded carbon atoms.

Don't forget that **primary alcohols** are oxidised by dilute sulphuric acid and potassium dichromate(VI) first to an aldehyde and then to a carboxylic acid. **Secondary alcohols** are oxidised to ketones and **tertiary alcohols** are not oxidised.

Alcohols don't react with nucleophiles such as CN^- ions (from KCN)

Don't use 'added up' formulae such as C_3H_7OH; this could be either propan-1-ol or propan-2-ol.

The product of a halogenoalkane with potassium (or sodium) hydroxide **depends on the solvent**. Aqueous conditions give an alcohol (nucleophilic substitution), ethanolic solutions an alkene (elimination).

Key points to remember

Definitions

- **Addition reaction**: a reaction in which two substances combine to form a **single** product.
- **Substitution reaction**: a reaction in which one atom or group replaces another in a molecule.
- **Elimination reaction**: a reaction in which a simple molecule, such as HBr or H_2O, is removed from a larger one and is not replaced by any other atom or group. The result is a compound with a double bond.
- **Free radical**: a species with an unpaired electron
- **Nucleophile**: a species with a lone pair of electrons, which it donates to form a covalent bond. It is attracted to electron deficient sites
- **Electrophile**: a species that accepts a lone pair of electrons and forms a new covalent bond. It is attracted to electron rich sites

Reactions of alkanes e.g. ethane

Reaction	Conditions
$+O_2$: $2C_2H_6 + 7O_2 \rightarrow 4CO_2 + 6H_2O$	Heat or spark
$+Cl_2$(or Br_2): $C_2H_6 + Cl_2 \rightarrow C_2H_5Cl + HCl$	UV light or sunlight

Key points to remember

Reactions of alkenes e.g. ethene

Reaction	**Conditions / observations**
$+Br_2(l)$ or in hexane $CH_2{=}CH_2 + Br_2 \rightarrow CH_2BrCH_2Br$	Red-brown bromine turns colourless
$+Br_2$ (aq) $CH_2{=}CH_2 + Br_2 + H_2O \rightarrow CH_2BrCH_2OH + HBr$	Red-brown bromine water turns colourless
+HBr (or HCl or HI) $CH_2{=}CH_2 + HBr \rightarrow CH_3CH_2Br$	Mix gases at room temperature
$+H_2$ $CH_2{=}CH_2 + H_2 \rightarrow CH_3CH_3$	Pass gases over nickel catalyst at 100 °C or over platinum catalyst at room temperature
$+H_2O$ $CH_2{=}CH_2 + H_2O(g) \rightarrow CH_3CH_2OH$	Pass ethene and steam over a phosphoric(V) acid catalyst on silica at 300 °C and a pressure of 60 atm.
polymerisation $n\ CH_2{=}CH_2 \rightarrow \text{-}[CH_2{-}CH_2]\text{-}_n$	Either: 1000 atm pressure, 150 °C and a trace of oxygen. Or: triethyl aluminium and titanium tetrachloride catalysts

Summary of the reactions of alkenes

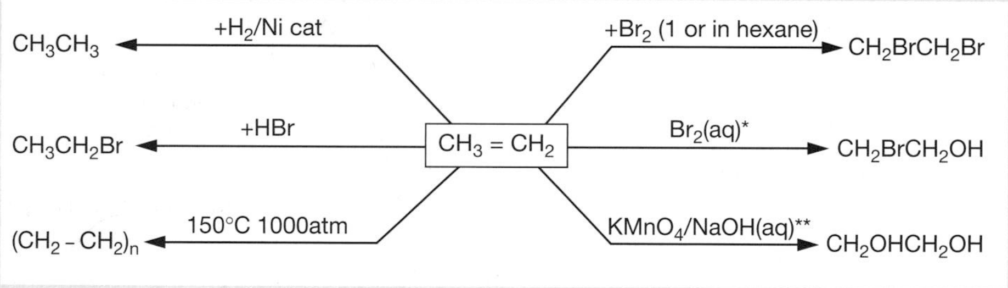

* Not required by Edexcel

** Only required by OCR

Reactions of alcohols

Reaction	**Conditions / observations**
+ potassium dichromate(VI) and dilute sulphuric acid $CH_3CH_2OH + 2[O] \rightarrow CH_3COOH + H_2O$ $CH_3CH(OH)CH_3 + [O] \rightarrow CH_3COCH_3 + H_2O$ $(CH_3)_3COH \rightarrow$ no reaction	Heat under reflux Primary alcohols are oxidised via aldehydes to carboxylic acids and turn the orange dichromate green Secondary alcohols are oxidised to ketones and also turn the orange dichromate green Tertiary alcohols are not oxidised and the dichromate stays orange
+ concentrated sulphuric acid $CH_3CH_2OH - H_2O \rightarrow CH_2{=}CH_2$	Heat to 170 °C with excess sulphuric acid and the alcohol will be dehydrated (or pass ethanol vapour over hot Al_2O_3)

Alcohols

- **Primary alcohols** have one carbon atom directly bonded to the carbon of the C—OH group:

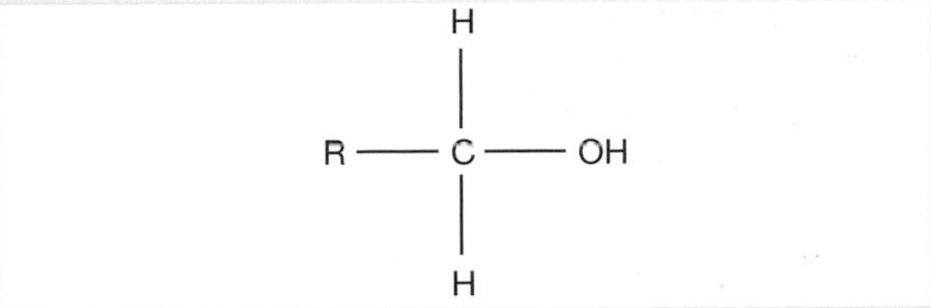

- **Secondary alcohols** have two carbon atoms directly bonded to the carbon of the C—OH group:

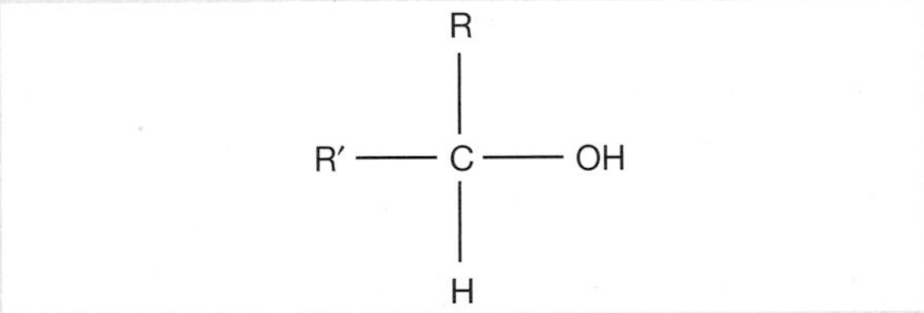

- **Tertiary alcohols** have three carbon atoms directly bonded to the carbon of the C—OH group:

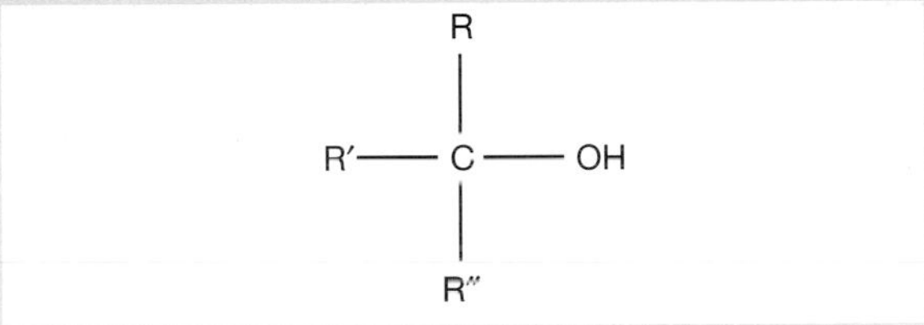

Summary of the reactions of alcohols

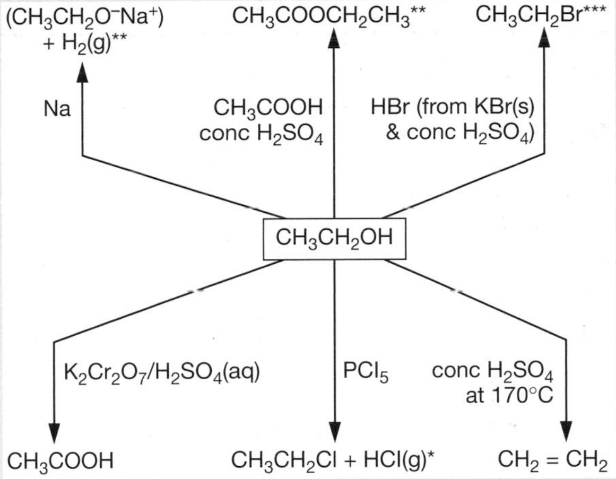

* Edexcel only ** OCR only
*** Edexcel and OCR

Summary of the reactions of halogenoalkanes

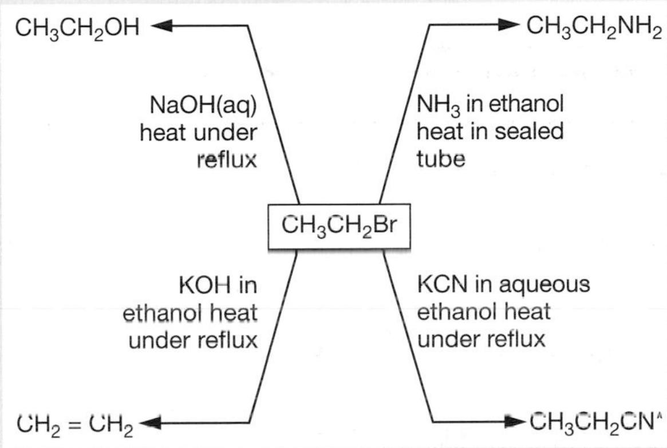

* not required by OCR

Reactions of halogenoalkanes (haloalkanes)

Reaction	Conditions
+ **aqueous** NaOH (or KOH) $CH_3CH_2Br + OH^- \rightarrow CH_3CH_2OH + Br^-$	Heat under reflux in aqueous solution. (a substitution reaction)
+ KOH dissolved in **ethanol** $CH_3CH_2Br + KOH \rightarrow CH_2{=}CH_2 + H_2O + KBr$	Heat under reflux in ethanolic solution (an elimination reaction)
+ NH_3 $CH_3CH_2Br + NH_3 \rightarrow CH_3CH_2NH_2 + HBr$ or to the salt $\rightarrow CH_3CH_2NH_3^+Br^-$	Heat the ammonia dissolved in ethanol with the bromoethane in a sealed tube
+ KCN (not required for OCR candidates) $CH_3CH_2Br + KCN \rightarrow CH_3CH_2CN + KBr$	Heat under reflux in an aqueous ethanolic solution

Q1

Consider the reaction scheme below:

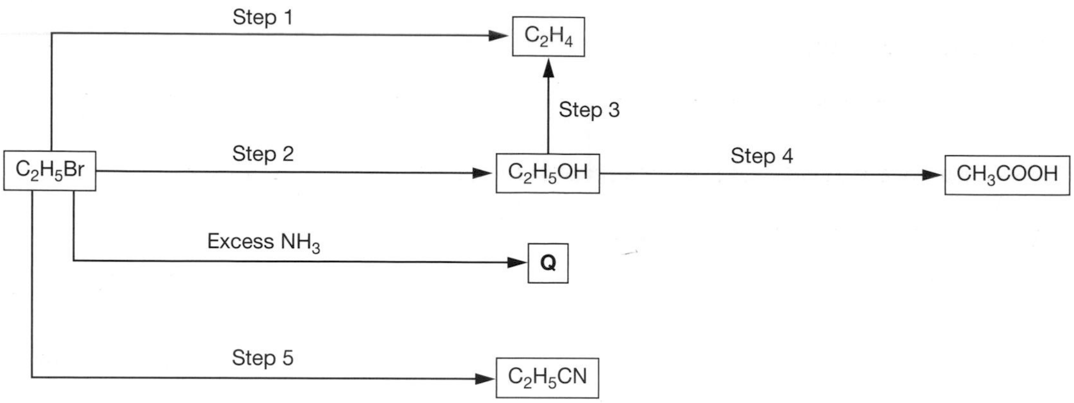

(a) State the reagents and conditions for:

(i) step 1

Reagent: .... Conditions: ....

[3 marks]

(ii) step 2

Reagent: .... Conditions: ....

[2 marks]

(iii) step 3

Reagent: .... Conditions: ....

[2 marks]

(iv) step 4

Reagents: .... Conditions: ....

[2 marks]

(v) step 5 (not OCR candidates)

Reagent: .... Conditions: ....

[3 marks]

(b) What would you observe when carrying out step 4?

....

[2 marks]

(c) Write the full (displayed) structural formula of Q.

[1 mark]

(d) Bromoethane, C_2H_5Br, could be made from ethane.

(i) State the reagent and conditions of this reaction.

Reagent: .... Conditions:..

[2 marks]

(ii) Classify this reaction.

....

[2 marks]

(iii) Why is this not a good method for preparing bromoethane from ethane?

....

....

[1 mark]

[Total 20 marks]

Q2

(a) Compound X is a mixture of two cis/trans (geometric) isomers. Its composition by mass is C = 47.1%, H = 6.5% and Cl = 46.4%.

(i) Calculate the empirical formula of X.

[3 marks]

(ii) The molecular formula is the same as the empirical formula. Draw diagrams showing the two cis/trans isomers of substance X.

[2 marks]

(iii) There are two non-cyclic structural isomers of X. Draw them.

[2 marks]

(b) Halogenoalkanes such as bromoethane, C_2H_5Br, undergo many nucleophilic substitution reactions.

(i) Explain the meaning of the term nucleophile.

....

....

[2 marks]

(ii) Give an example of a nucleophile that will react with bromoethane.

....

[1 mark]

(iii) Explain the meaning of the term substitution.

....

....

[1 mark]

[Total 11 marks]

Q3

(not for Edexcel candidates)

Chlorine reacts readily with ethene in the dark but does not react with methane unless sunlight or another source of ultraviolet light is present.

State and describe, with the aid of suitable equations:

- The mechanism of the reaction between chlorine and methane.
- The mechanism of the reaction between chlorine and ethene.
- The type of fission that the Cl—Cl bond undergoes in each of the mechanisms.

....

....

....

....

....

....

....

....

....

....

....

....

[12 marks]

[Total 12 marks]

The answers to these questions are on pages 89 and 90.

10 Answers to Questions to try

Notes: A tick against part of an answer indicates where a mark would be awarded.

Chapter 1 Atomic Structure

Q1 How to score full marks

(a) (i) $1s^2, 2s^2, 2p^3$. ✓

(ii) The highest occupied energy level is a p orbital. ✓

(iii) $N(g) \rightarrow N^+(g) + e^-$. ✓ ✓

(iv) N is $1s^2, 2s^2, 2p_x^1 2p_y^1 2p_z^1$ whereas O is $1s^2, 2s^2, 2p_x^2 2p_y^1 2p_z^1$. ✓

The **repulsion** of to 2 electrons in the $2p_x$ orbital of **oxygen** makes it **easier** to remove one of them. ✓

Although F is $1s^2, 2s^2, 2p_x^2 2p_y^2 2p_z^1$, the nuclear charge is **two more** than in nitrogen ✓ and there are **no** extra shielding (inner shell) electrons, and so the $2p_x$ electron in fluorine is held more firmly, and so is harder to remove. ✓

(b) (i) Mass number is the sum of the number of protons and neutrons ✓ in an atom ✓ (of the isotope).

Relative atomic mass is the average ✓ mass of the naturally occurring isotopes ✓ (taking into account their relative abundance) of the different atoms of the element relative to (or divided by) $\frac{1}{12}$th the mass of a carbon 12 atom. ✓

(ii) % of ^{121}Sb – 60 and % ^{123}Sb – 40. ✓

A_r of Sb = $(121 \times 60 + 123 \times 40) \div 100 = 121.8$ ✓

(iii) The ions are accelerated by an electric field ✓

They are deflected by a magnetic field. ✓

The heavier ions are deflected the least. ✓

Examiner's comments

(a) (ii) Do not say "partially filled p orbital"

(iii) Remember that ionisation energy refers to **gaseous atoms.**

(iv) To gain full marks you have to mention, where appropriate

The change in nuclear charge

The change in shielding (if any)

The repulsion due to two electrons in the same orbital

The size of the atom.

Here the last point is not very important, as the atoms are all of a very similar size.

(b) (i) As the question asked to **distinguish** the two terms, you must make it clear that the mass number refers to a particular isotope of an element, whereas relative atomic mass refers to the natural mixture of the isotopes of an element.

(ii) The four parts to a mass spectrometer are

Ionisation (of a gaseous sample)

Acceleration (by a electric field)

Deflection (by a magnetic field)

Detection (not asked for in this question).

Q2 How to score full marks

(a) The relative masses and charges are:

	Relative mass	Relative charge
Proton	1	+1
Neutron	1	0
Electron	1/1860 ✓	–1 ✓

Isotopes have ✓ the same number of protons ✓ and hence electrons, but a different number of neutrons. ✓ This means that the masses ✓ of each isotope differ from each other.

Chemical properties are determined by the electronic configuration ✓ of the atoms of the element. As isotopes have the same number of protons, they have the same number of electrons and the same ✓ electronic configuration. This results in them having identical chemical properties.

(b) The element is vapourised by heating it ✓ in a vacuum ✓. It is then bombarded by high-speed electrons ✓, which knock off an electron and form positive ions ✓. These are then accelerated ✓ by a potential difference. The beam of positive ions is then deflected by a powerful electromagnet ✓. Finally the ions are detected ✓.

The heavier ions are deflected less ✓ and so are separated from the lighter ions of the element.

Examiner's comments

(a) Do not say that the relative mass of an electron is zero. You would be allowed to state that it is negligible or almost zero. You must state in what way the atomic structures are the same and how they differ as well as the resulting difference in physical property (mass). Lastly you must make it clear that you understand that it is the electronic configuration that determines the chemical properties.

(b) It might help to sketch a labelled diagram of a mass spectrometer. In this way, if you have omitted something in your description, it might score from the diagram. Don't forget that the spectrometer only works when all the air has been removed and that most elements need to be heated to get them into the gaseous state.

Chapter 2 Formulae, equations and amount of substance

Q1 How to score full marks

(a) **(i)** moles of Ca $= \dfrac{\text{mass}}{A_r} = \dfrac{0.122}{40} = 0.00305$ mol ✓

moles of HCl $=$ concentration × vol $= 1.23 \times \dfrac{25}{1000} = 0.03075$ mol ✓

0.00305 mol Ca requires $\frac{2}{1} \times 0.00305 = 0.0061$ mol HCl ✓

$0.0061 < 0.03075$, therefore HCl is in excess. ✓

(ii) moles of $H_2 = 0.00305 \times \frac{1}{1} = 0.00305$ mol ✓

volume of H_2 gas = mol × molar volume = 0.00305 × 24 000

= 73.2 cm^3. ✓

(b) purity = $\frac{\text{actual volume of hydrogen}}{\text{theoretical volume of hydrogen}} \times 100$ ✓ $= \frac{70.1}{73.2} \times 100 = 95.8\%$ ✓

(c) (i)

	%	÷	A_r ✓		÷ smallest ✓	
Na	34.3	÷	23	= 1.49	÷ 1.49 =	1
C	17.9	÷	12	= 1.49	÷ 1.49 =	1
O	47.8	÷	16	= 2.99	÷ 1.49 =	2

Empirical formula is $NaCO_2$ ✓

(ii) Formula mass of $NaCO_2$ = 23 + 12 + 32 = 67 ✓

$\frac{134}{67}$ = 2. Therefore molecular formula is $Na_2C_2O_4$ ✓

Examiner's comments

Note how the answers have been set out showing all the working.

This allows partial marks to be awarded if a careless mistake or an error is made.

In **(ii)** the question asked for the molecular formula to be **deduced**. This means that, however simple the calculation, some working has to be given.

Q2 How to score full marks

(a) Concentration = moles ÷ volume = 0.732 mol ÷ (250/1000) dm^3 = 2.93 ✓ mol dm^{-3} ✓

(b) (i) moles H_2SO_4 = concentration × volume = 1.24 mol dm^{-3} × (25.0/1000) dm^3 = 0.0310 mol ✓

moles NH_3 = 2 × moles H_2SO_4 = 2 × 0.0310 = 0.0620 mol ✓

concentration = moles ÷ volume = 0.0620 mol ÷ (30.8/1000) dm^3 = 2.01 mol dm^{-3} ✓

(ii) moles $(NH_4)_2SO_4$ = moles H_2SO_4 = 0.0310 mol ✓

molar mass $(NH_4)_2SO_4$ = 2 × (14 + 4) + 32 + (4 × 16) = 132 g mol^{-1} ✓

mass $(NH_4)_2SO_4$ = moles × molar mass = 0.0310 mol × 132 g mol^{-1} = 4.09 g ✓

(c) moles of NaCl = mass / molar mass = 800 g / (23 + 35.5) g mol^{-1} = 13.7 mol ✓

2 mol NaCl are needed to make 1 mol Na_2CO_3

mol Na_2CO_3 = $\frac{1}{2}$ × 13.7 = 6.85 mol ✓

molar mass Na_2CO_3 = (2 × 23) + 12 + (3 × 16) = 106 g mol^{-1} ✓

mass Na_2CO_3 = moles × molar mass = 6.85 mol × 106 g mol^{-1} = 726 g ✓ (allow 725 g)

(d) (i) The equation is:

$$Mg_3N_2 + 6H_2O \rightarrow 3Mg(OH)_2 + 2NH_3$$ ✓

(ii) moles NH_3 = mass / molar mass = 0.263 g / (14 + 3) g mol^{-1} = 0.0155 mol ✓

number of molecules = moles × Avogadro's constant = 0.0155 mol × 6.02 × 10^{23}
= 9.33 × 10^{21} molecules ✓ (allow 9.31 × 10^{21})

Examiner's comments

(a) The appropriate unit of concentration is mol dm^{-3}, so don't forget to convert the volume in cm^3 to a volume in dm^3.

(b) (i) The ratio of NH_3 to H_2SO_4 is 2:1, therefore moles of NH_3 = 2 x moles H_2SO_4

(ii) Even though the stoichiometry is 1:1, you must state that moles of $(NH_4)_2SO_4$ equals moles of H_2SO_4

(c) Do take great care with converting moles of one substance to moles of another. This time the stoichiometry requires the moles of NaCl to be halved.

(d) (i) Balance the element that appears in the most substances last. In this equation that is the hydrogen.

Q3 How to score full marks

(a) $Al_2O_3 + 6HCl \rightarrow 2AlCl_3 + 3H_2O$ ✓

(b) $4LiNO_3 \rightarrow 2Li_2O + 4NO_2 + O_2$ ✓

(c) $2K + 2H_2O \rightarrow 2KOH + H_2$ ✓

(d) $3Ca(OH)_2 + 2Na_3PO_4 \rightarrow Ca_3(PO_4)_2 + 6NaOH$ ✓

(e) $2Fe^{3+}(aq) + Sn^{2+}(aq) \rightarrow 2Fe^{2+}(aq) + Sn^{4+}(aq)$ ✓

(f) $5I^-(aq) + IO_3^-(aq) + 6H^+(aq) \rightarrow 3I_2(aq) + 3H_2O(l)$ ✓ ✓

Examiner's comments

(a) 2Al in Al_2O_3 requires $2AlCl_3$ on right and hence 6Cl and so 6HCl on left

(d) Either count all the atoms of each element or count OH groups and PO_4 groups separately. $2PO_4$ groups on right, therefore there must be 2 on the left, therefore 6Na's required (in $2Na_3PO_4$), so 6NaOH on right.

(e) This equation must balance for charge. $2x3^+ + 2^+ = 8^+$ on left and $2x2^+ + 4^+ = 8^+$ on right.

(f) 3 oxygen atoms in IO_3^-, therefore $3H_2O$ on right. This means $6H^+$ on left. As the right is neutral there must be $5I^-$ on the left to make it neutral too. 5^- & 1^- & $6^+ = 0$ (neutral).

Chapter 3 Structure and bonding

Q1 How to score full marks

The shape of a molecule is determined according to VSEPR theory, which states that the electron pairs in the outer orbit around the central atom of a molecule repel each other and so take up a shape of maximum separation of the electron pairs. ✓

When predicting the shape, the number of bond and lone pairs of electrons must be determined. ✓ This number tells you the arrangement of the electron pairs around the central atom. For instance if there are four pairs, they are arranged tetrahedrally. If there are no lone pairs, the shape of the molecule is the same as the electron arrangement.

Examples:

Boron trichloride, BCl_3, has ✓ three bond pairs and no lone pairs around the boron atom and so will be trigonal planar, ✓ with a bond angle of 120°. ✓

Methane, CH_4, has four bond pairs ✓ and no lone pairs around the carbon atom and so the shape of the molecule is tetrahedral ✓ with a H–C–H bond angle of 109½°. ✓

Effect of lone pairs:

The presence of lone pairs alters the shape. If an atom is surrounded by three bond pairs and one lone pair, as in ammonia, NH_3, the electron pairs will ✓ be arranged in a tetrahedron, but the three hydrogen atoms will be arranged in a pyramid ✓ around the nitrogen atom. Because the repulsion between a lone pair and a bond pair is greater than that between two bond pairs, the bond angle is reduced from 109½° to 105°. ✓

Water, H_2O, has ✓ two bond and two lone pairs around the oxygen atom, and so the molecule will have a V shape ✓ with the bond angle reduced further by the lone pair/lone pair repulsion to 104½°. ✓

Examiner's comments

The most important point is that shape is determined by electron pair repulsion, not by the repulsion of bonds or atoms.

The total number of bond and lone pairs determines the arrangement of the electron pairs but, as lone pairs cannot be seen, the shape is that taken up by the central atom and those atoms surrounding it. Thus ammonia has the electron pairs arranged tetrahedrally, but the lone pair is above the nitrogen with the three hydrogen atoms arranged pyramidally beneath.

Q2 How to score full marks

(a) The hydrogen bond is an **intermolecular** force which in water is between a δ+ hydrogen atom in one molecule and a δ– oxygen atom in another molecule.

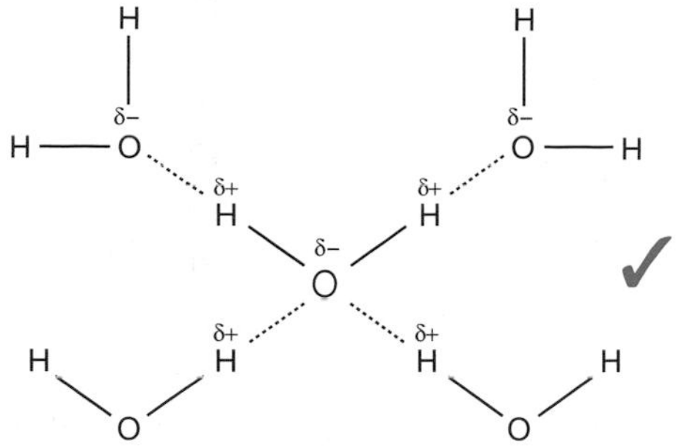

(b) There are **hydrogen** bonds between HF molecules but not between HCl molecules. Therefore the **intermolecular** forces are stronger in HF than in HCl, and so HF has a higher boiling point.

In HCl and HBr the main intermolecular forces are instantaneous induced dipole/induced dipole (also called London or dispersion) forces.

HCl has fewer electrons (18) then does HBr (36 electrons), and so the instantaneous induced dipole/induced dipole (dispersion) forces in HCl are weaker than those in HBr, and hence HCl has the lower boiling point.

> **Examiner's comments**
>
> **(a)** You must draw a diagram showing the difference between the covalent bonds and the hydrogen bonds. Note that each oxygen atom in water forms **two** covalent bonds and **two** hydrogen bonds, and each hydrogen **one** covalent and **one** hydrogen bond.
>
> **(b)** Hydrogen bonds only occur between hydrogen and **F**, **O** or **N** atoms.
>
> HCl is more polar than HBr, but the stronger instantaneous induced dipole/induced dipole (dispersion) forces between HBr molecules outweigh the slightly stronger dipole/dipole forces between HCl molecules.

Q3 How to score full marks

(a) **(i)** The property which causes the bonds to be polar is the difference in electronegativity of the two atoms in the bond.

(ii) Hydrogen bonding occurs between HF molecules but this does not happen with HCl (as the Cl atom is too big and is less δ– than the F atom). With HCl the intermolecular forces are van der Waals forces (instantaneous induced dipole and also permanent dipole/dipole forces), which are weaker than hydrogen bonds. Thus more energy is required to separate HF molecules than to separate HCl molecules, therefore HF has the higher boiling point.

(b) One of the lone pairs of electrons on the Cl^- ion is donated into the empty $3p_z$ orbital of the aluminium forming the bond.

(c)

PCl_5

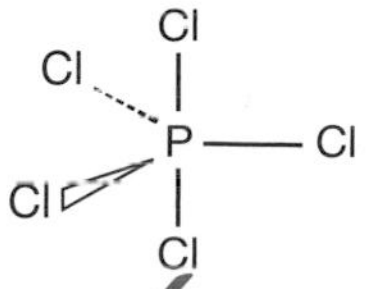

Bond angles are 120° and 90°.

PCl_4^+

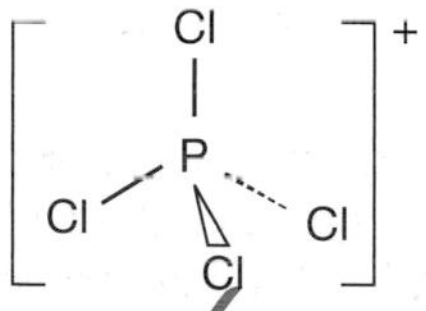

Bond angle is 109½°.

Examiner's comments

(a) You must state the type of intermolecular force in both, and which is the stronger, and link strength of force to energy required to separate the molecules. Remember that covalent bonds are not broken, and so the strength of the H–F and H–Cl covalent bonds is nothing to do with the question.

(b) Don't forget to state how the aluminium can accept the two electrons. In $AlCl_3$, the aluminium has only six electrons in its outer orbit.

(c) The Cl–P–Cl bond angles in the plane are 120°, and those between the top Cl and those in the plane are 90°.

Q4 How to score full marks

(a) In sodium metal there is a regular lattice of Na^+ ions ✓ attracting the delocalised ✓ electrons. These electrons are not in a fixed place and are mobile ✓, so they are able to move through the solid and so can conduct electricity.

(b) The Na^+ and Cl^- ions are arranged in a regular lattice, with six Na^+ ions around each Cl^- ion and six Cl^- ions around each Na^+ ion ✓ There is a strong electrostatic force holding the ions together ✓ The ions are fixed and so cannot move ✓ through the solid and so it does not conduct electricity.

(c) (i) As the solid sodium chloride is heated, the ions vibrate more ✓ until the vibrations become so strong that, at the melting point, the lattice begins to break down ✓ and the ions become free to move about randomly in the molten liquid.

(ii) As the ions are free to move ✓ the molten sodium chloride will conduct electricity.

Examiner's comments

(a) Electricity is the flow of charged particles. In a solid metal the electrons are not fixed in any position and are able to move between the much larger positive ions and so conduct electricity.

(b) All the electrons are localised on the ions and the ions themselves are fixed in the lattice and are unable to move about.

(c) Don't say that the ions **start** to vibrate when heated; they are already are vibrating. The lattice structure is only broken down at the melting point, and then the ions are no longer fixed. They move when a potential difference is applied and so conduct electricity.
Remember that metals conduct by movement of electrons, and ionic substances conduct, when molten or in solution, by movement of ions.

Chapter 4 The Periodic Table and Group II (Group 2)

Q1 How to score full marks

(a) (i) Silicon has a giant atomic structure in which each silicon atom is covalently bonded to four others ✓ These strong covalent bonds have to be broken ✓ to melt silicon and this requires a large amount of energy, and so its melting point is very high.

Sulphur and chlorine are both simple molecular substances ✓ with instantaneous induced dipole (dispersion) forces ✓ between the molecules. The strength of these forces depends upon the total number of electrons ✓ in the molecule. Sulphur is S_8 and has $8 \times 16 = 128$ electrons, whereas chlorine is Cl_2 and only has $2 \times 17 = 34$ electrons ✓ Thus the intermolecular forces in sulphur are stronger and more energy ✓ is required to separate the molecules.

(ii) The electronic configurations are:

P: $1s^2, 2s^2\ 2p^6, 3s^2\ 3p_x^{\ 1}\ 3p_y^{\ 1}\ 3p_z^{\ 1}$ and S: $1s^2, 2s^2\ 2p^6, 3s^2\ 3p_x^{\ 2}\ 3p_y^{\ 1}\ 3p_z^{\ 1}$ ✓

Phosphorus has to lose an electron from a fairly stable configuration in which an energy level (the 3p) is half filled ✓ but sulphur loses one of the $3p_x$ pair of electrons. The two electrons in the same orbital repel each other ✓ and so less energy is required to remove one of them.

(b) Magnesium burns with a bright white flame when heated in steam.

$$Mg + H_2O \rightarrow MgO + H_2$$

When calcium is added to cold water, bubbles of gas are rapidly given off.

$$Ca + 2H_2O \rightarrow Ca(OH)_2 + H_2$$

Examiner's comments

(a) (i) Melting giant atomic structures such as silicon and graphite are the only occasions when covalent bonds need to be broken for melting to happen. Dispersion forces depend upon the number of electrons and **not** on the mass of the molecule.

(ii) To explain the difference in ionisation energies, the electron configuration must be given in detail with the number of electrons in each of the three 3p orbitals shown.

(b) Beryllium does not react even when heated in steam. Magnesium reacts with steam to form its oxide and hydrogen, and the other Group II metals react with liquid water to form hydroxides and hydrogen. Do not give as your observations that **hydrogen** is given off. What you see are bubbles of gas. That it is hydrogen is a deduction not an observation.

Q2 How to score full marks

(a) The 1st ionisation energy decreases because the radius increases.

The extra nuclear charge is compensated for by the equivalent increase in inner (shielding) electrons.

(b) (i) $Ca(s) + 2H_2O(l) \rightarrow Ca^{2+}(aq) + 2OH^-(aq) + H_2(g)$

(ii) $Mg(s) + H_2O(g) \rightarrow MgO(s) + H_2(g)$

(iii) The 1st and 2nd ionisation energies get less so it gets easier to remove 2 electrons to form a 2^+ ion.

(c) (i) Radium reacts very rapidly
to form radium hydroxide and hydrogen
(or the equation $Ra + 2H_2O \rightarrow Ra(OH)_2 + H_2$).

(ii) Insoluble.

(iii) Any value between 218 and 221 pm.

Examiner's comments

(a) There are 3 marks for the question, so 3 points have to be made.

(b) (i) Calcium hydroxide is ionic and (reasonably) soluble so the Ca^{2+} and the OH^- ions have to be written separately. The other substances are not ionic.

(iii) Group II metals form 2+ ions, and so mention must be made of both the 1st and 2nd ionisation energies.

(c) (i) The prediction must state what happens and how quickly.

Chapter 5 Redox and Group VII (Group 7)

Q1 How to score full marks

(a) (i) Oxidation number is the charge that an element would have if all the bonds in the substance were ionic.

(ii) The oxidation number of chlorine in Cl_2 is zero, in OCl^- is +1 and in $NaClO_3$ it is +5.

(b) (i) $Cl_2 + H_2O \rightleftharpoons HCl + HOCl$

$Cl_2 + 2OH^- \rightarrow Cl^- + OCl^- + H_2O$

(ii) The chlorine has been both oxidised and reduced (disproportionated). ✓

Its oxidation number has simultaneously gone up form zero in Cl_2 to +1 in H_2O and OCl^- and down from zero in Cl_2 to –1 in HCL and Cl^-. ✓

(c) **(i)** $FeO_4^- + 8H^+ + 4e^- \rightarrow Fe^{3+} + 4H_2O$ ✓✓

(ii) $Fe^{2+} \rightarrow Fe^{3+} + e^-$ ✓

(iii) $FeO_4^- + 8H^+ + 4Fe^{2+} \rightarrow 5Fe^{3+} + 4H_2O$ ✓✓

Examiner's comments

(a) (ii) The oxidation number of an uncombined element is zero. In OCl^-, –2 + Cl = –1 and in $NaClO_3$, +1 + Cl + (–6) = 0.

(c) (i) Fe is +7 in FeO_4^-, and iron's oxidation number changes by –4, so $4e^-$ must be on the left. $8H^+$ ions are needed to form $4H_2O$ from the 4 oxygens in FeO_4^-.

(iii) Equation (ii) has to be multiplied by 4, so that both equations have the same number of electrons.

Q2 How to score full marks

Examiner's comments

(a) First you must state the trend. Then you must identify the type of substance (ionic, simple covalent, hydrogen bonded covalent). Then describe the type of intermolecular force, showing how the strength varies down the group. Finally link the strength to the energy needed for separation (boiling). Don't fall into the trap of discussing the strength of covalent bonds. They are **not** broken in boiling.

(b) Make sure that the answers for both substances are divided into two parts: adding silver nitrate and then adding ammonia.

(c) As in **(a)** you must first state the trend. Be careful that you don't confuse a halogen (chlorine) with a halide ion (Cl^-). It is vital not only to state what happens when a higher halogen is added, but that nothing happens when a lower halogen is used. You must link the displacement reaction to redox. Finally make sure that you state the initial colour as well as the final colour, when describing the observations.

(a) The boiling points of the halogens increase down the group from fluorine to iodine. ✓

All the halogens form simple molecules of formula X_2, ✓ and the forces between these molecules are instantaneous induced dipole (dispersion) forces. ✓ The strength of these forces depends upon the total number of electrons in the molecule. ✓ F_2 contains only 18 electrons, Cl_2 34 and so on to I_2 which contains 106. Thus the strength of the intermolecular forces increases down the group from fluorine to iodine ✓ and so the amount of energy required to separate the molecules on boiling increases. ✓ This is the reason why the boiling points steadily increase.

(b) Sodium chloride. When silver nitrate is added the colourless solution would give a white precipitate. ✓

A colourless solution ✓ would form when aqueous dilute ammonia was added.

Sodium bromide. The colourless solution would give a pale cream coloured precipitate ✓ on addition of the silver nitrate.

No change ✓ would be observed when dilute aqueous ammonia was added.

(c) The oxidising power of the halogens decreases down the group. ✓

This is shown by the reaction in which a halogen higher in the group will displace a halogen lower from its halide salt in solution, ✓ but a lower halogen will not ✓ displace a higher one from its halide salt in solution.

For example chlorine will displace iodine from aqueous potassium iodide, but iodine will not displace chlorine from aqueous potassium chloride.

The reaction of chlorine with potassium iodide is shown by the ionic equation:

$$Cl_2(g) + 2I^-(aq) \rightarrow I_2(s) + 2Cl^-(aq)$$ ✓

The colourless solution of potassium iodide will change to a grey precipitate of iodine ✓. The chlorine atoms in Cl_2 are reduced because each gains one electron from an I^- ion which becomes oxidised ✓.

When a brown solution of iodine ✓ is added to colourless aqueous potassium chloride, the brown colour of the I_2 remains. As there is no reaction, there is no equation.

Chapter 6 Energetics

Q1 How to score full marks

Examiner's comments

(a) (ii) it is a good idea to give an equation with state symbols as part of an answer to a definition in energetics.

(b) The first two steps of the cycle are the reverse of formation and so equals $-\Delta H_f$

(c) (i) One O—H bond in water breaks. The H bonds to one carbon atom and the O in the OH bonds to the other carbon atom

(a) **(i)** Standard conditions are a constant pressure of 101 kPa (1 atm), ✓
a stated temperature (usually 298K) ✓
and any solutions at a concentration of 1 mol dm^{-3}. ✓

(ii) It is the enthalpy change when 1 mole of the substance ✓
is formed from its elements under standard conditions. ✓
For ethene it is the enthalpy change for the reaction:
$2C(s) + 2H_2(g) \rightarrow CH_2{=}CH_2(g)$ ✓

(b)

$$CH_2{=}CH_2(g) + H_2O(g) \rightarrow CH_3CH_2OH(g)$$ ✓

$$2C(s) + 3H_2(g) + \tfrac{1}{2}O_2(g)$$

$\Delta H_r^\ominus = -\Delta H_f^\ominus(\text{ethene}) - \Delta H_f^\ominus(\text{water}) + \Delta H_f^\ominus(\text{ethanol})$ ✓
$= -(+53) - (-242) + (-235) = -46 \text{ kJ mol}^{-1}$ ✓

(c) **(i)**

Bonds broken (endothermic)	Bonds made (exothermic)
C═C = + 612	C—C = − 348
H—O = + 463	C—H = − 412
	C—O = − 360
+ 1075 kJ ✓	− 1120 kJ ✓

$\Delta H_r^\ominus = +1075 + (-1120) = -45 \text{ kJ mol}^{-1}$ ✓

(ii) The bond energies in the data are average values, ✓
so there is a slight difference between the average bond enthalpies and the actual bond enthalpies in these compounds. ✓

Q2 How to score full marks

Examiner's comments

(a) Straight lines need to be drawn for the period up to 3 minutes and from 5 to 10 minutes. These lines are then extrapolated to 3 minutes (when the reaction started) and ΔT measured at the 3-minute point.

ΔH is negative as the reaction is exothermic.

(b) The mass is the sum of the copper sulphate and the water

(c) The difficult part is deducing the cycle. The clue is the ΔH values that you have: the ΔH calculated in part (b), and that given in the data of (c).

(a)

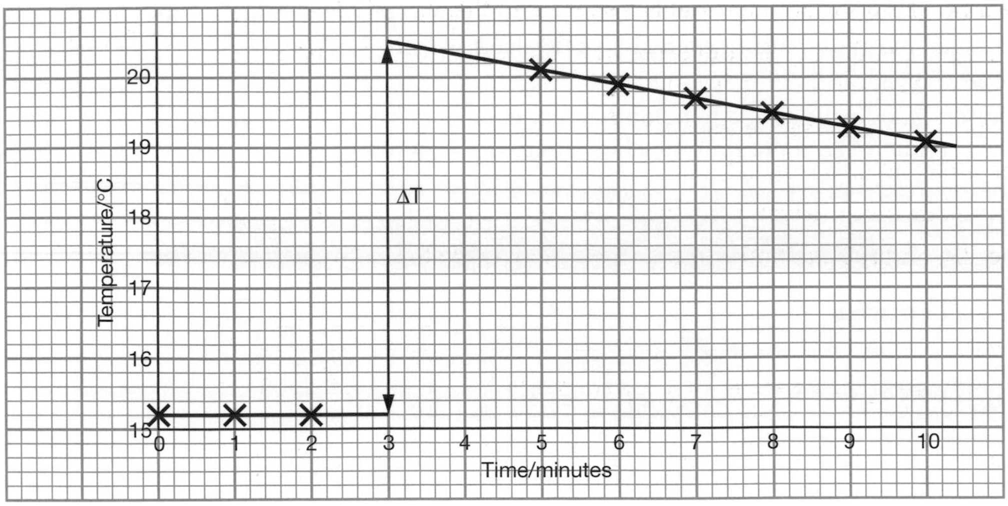

Axes labelled ✓ points correctly drawn ✓ lines drawn ✓ extrapolation to 3 minutes ✓
$\Delta T = 20.5 - 15.2 = 5.3\,°C$. ✓

(b) heat produced = $104.8 \times 4.18 \times 5.3 = 2322$ J = 2.322 kJ ✓

amount of anhydrous copper sulphate = $\dfrac{4.80\text{ g}}{159.5\text{ g mol}^{-1}}$ = 0.03009 mol ✓

heat produced per mole = $\dfrac{2.322}{0.03009}$ = 77.2 kJ mol^{-1} ✓

$\Delta H = -77.2$ kJ mol^{-1} ✓

(c) $CuSO_4(s) + 5H_2O(l) \rightarrow CuSO_4.5H_2O(s)$

↘ ↙ ✓

$CuSO_4(aq)$

or

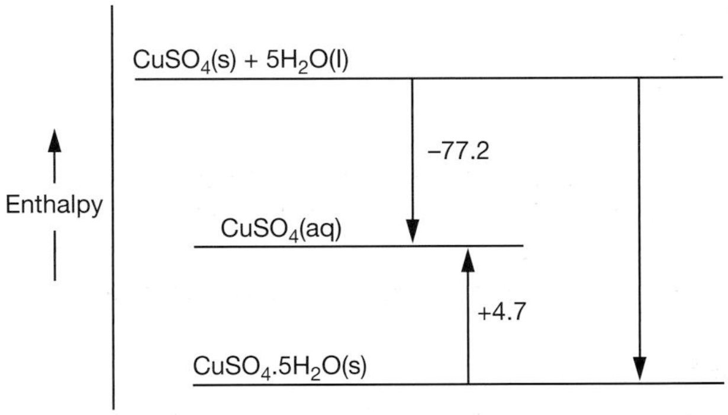

ΔH_r = enthalpy of solution of $CuSO_4(s)$ – enthalpy of solution of $CuSO_4.5H_2O$ ✓

$= -77.2 - (+4.7) = -81.9$ kJ mol^{-1} ✓

Chapter 7 Kinetics

Q1 How to score full marks

(a) The rate can be increased by:

an increase of temperature ✓
an increase the concentration of either or both reactants ✓
and the addition of a suitable catalyst. ✓

(b) The flour dust has a very high surface area and is dispersed in the air. ✓

The rate of collision between the oxygen molecules and the surface of the flour particles is very high, and so the reaction is rapid. ✓

In the home the flour is not dispersed in the air. ✓

(c) **(i)** None; there are no reactant gases. ✓

(ii) It slows down the rate because water will lower the concentration of the acid. ✓

(iii) None, because HBr and HCl are both strong acids, and it is the H^+ ions that react with the zinc.

> **Examiner's comments**
>
> **(a)** This reaction is carried out in solution, so pressure cannot alter the rate of reaction.
>
> **(b)** To score full marks, you need to say something about both the flour in the mill and in the home.
>
> **(c)** HCl is fully ionised into $H^+(aq)$ and $Cl^-(aq)$

Q2 How to score full marks

(a) The activation energy is the minimum energy that the molecules must have on collision in order to form the products. ✓✓

(b) **(i)**

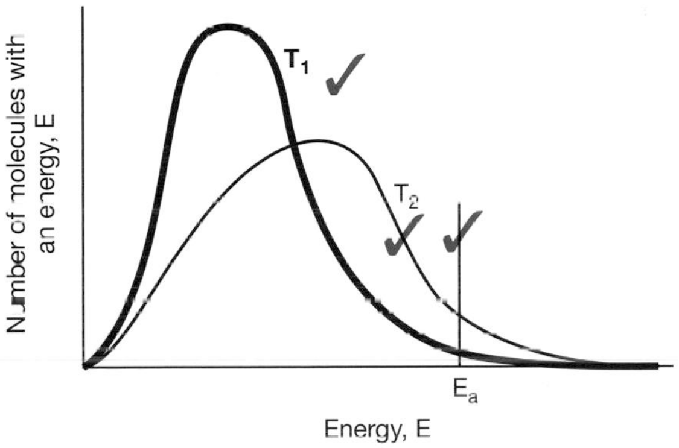

At the higher temperature the molecules have a higher average kinetic energy ✓ and so the area under the curve to the right of the E_a line is greater at T_2 than at T_1. ✓

This means that more molecules will have the activation energy on collision and so will react. ✓

This causes the reaction rate to increase. ✓

In addition there is a slight increase in rate due to the molecules moving faster and so colliding more often. ✓

(ii) The rate will increase. ✓

The reactants are gases and so an increase in pressure will result in their being more molecules in a given volume. ✓

This means that the frequency of collision will be increased ✓

> **Examiner's comments**
>
> **(a)** You need to draw two graphs at temperatures T_1 and T_2. These must be clearly labelled with the activation energy drawn in to the right of both peaks and will score three marks. The remaining five marks are for the explanation.
>
> **(b)** Firstly you must state whether the rate would be unchanged, increased or decreased. This will score 1 mark. There are 3 marks for this part, so you must make 2 more points to score full marks.

Q1 How to score full marks

(a)

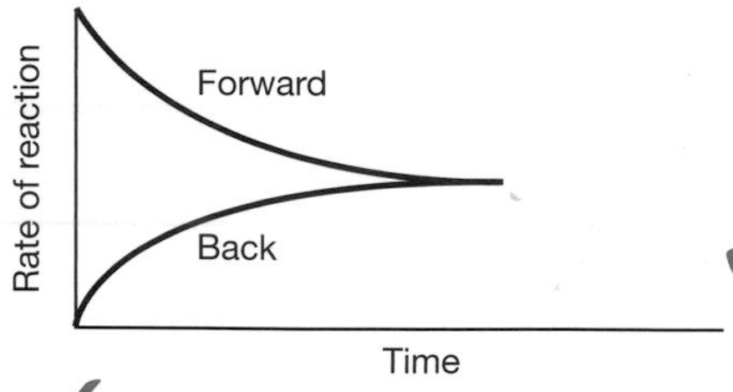

✓✓✓

(b) false ✓
true ✓
false ✓
true ✓

(c) (i) A temperature of 425 °C is chosen because:
as the reaction is exothermic left to right, a higher temperature would shift the position of equilibrium to the left which would result in a lower yield. ✓
a lower temperature would shift the equilibrium to the right causing the theoretical yield to increase, ✓
but the rate of reaction would be too low, and hence the time required to reach equilibrium would be too long. ✓
a compromise temperature of 425 °C is used, which is an economic balance between rate and yield. ✓

(ii) As there are fewer gas molecules on the right hand side, a high pressure would shift the position of equilibrium to the right, thus increasing the yield. ✓

But the yield is high at a pressure of 202 kPa, and so there is no need for expensive high pressure. ✓ (The pressure must be above atmospheric to push the chemicals through the plant.)

(iii) A catalyst speeds up the rate and so enables the reaction to take place quickly at a moderate temperature. ✓

It has no effect on the position of equilibrium or the theoretical yield. ✓

Thus it is more economic to run the plant at the same rate at a lower temperature with a catalyst, than without a catalyst when the reaction would require a higher temperature and hence produce a lower yield. ✓

(d) (i) The solution would turn yellow. ✓
The OH^- ions would react with the H^+ ions, thus lowering their concentration. ✓
The system adjusts to replace those H^+ ions by moving to the left. ✓

(ii) The solution would turn orange. ✓
The addition of sulphuric acid increases the concentration of H^+ ions. ✓
The system adjusts to remove them by moving to the right. ✓

Examiner's comments

(a) The rate of the reverse reaction is zero at the beginning. At equilibrium the two rates are equal.

(b) The equilibrium is dynamic and so both reactions are still taking place, but at equal rates.

(c) (i) You must explain the disadvantages of using either a higher temperature or a lower temperature.

(ii) At first sight a high pressure would seem to be a good idea, but high pressures are expensive and so are only used for reactions with low yields (such as the Haber process).

(d) You must state the colour change and then explain it in terms of equilibrium.

Chapter 9 Organic chemistry

Q1 How to score full marks

(a) (i) step 1 Reagent: potassium hydroxide (KOH)
Conditions: heat under reflux in ethanolic solution.

(ii) step 2 Reagent: aqueous sodium or potassium hydroxide
Conditions: heat under reflux

(iii) step 3 Reagent: concentrated sulphuric or phosphoric(V) acid.
Conditions: heat to 170 °C

Or Reagent: aluminium oxide, Al_2O_3, as catalyst
Conditions: pass ethanol vapour over hot aluminium oxide

(iv) step 4 Reagent: potassium dichromate(VI) + dilute sulphuric acid
Conditions: heat under reflux

(v) step 5 Reagent: potassium cyanide
Conditions: heat under reflux in an aqueous ethanol solvent

(b) The solution would go from orange to green.

(c)

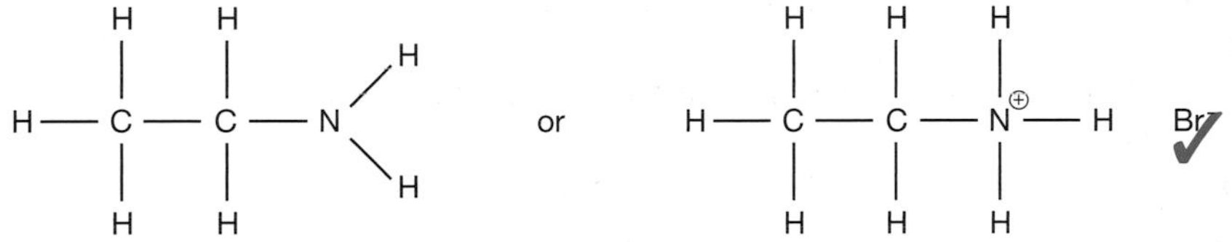

(d) (i) Reagent: bromine Conditions: UV (or sun) light

(ii) Free radical substitution

(iii) Some $C_2H_4Br_2$ and further substitution products would also be produced.

Examiner's comments

(a) Step 1 is elimination and so an ethanolic solution must be used, whereas step 2 is hydrolysis and aqueous alkali must be used.
Step 3 is dehydration and so the reagent is concentrated sulphuric acid or aluminium oxide.
Step 4 is oxidation. If heated under reflux, an acid is produced (but if the product is allowed to distill off as it is made, the aldehyde will be obtained).

(b) Do not forget to give the colour before and after reaction.

(c) Both the free amine and its salt can be given as answers.

Q2 How to score full marks

Examiner's comments

(a) (i) Do not round down to a whole number after dividing by the relative atomic mass.

(ii) The question told you that **X** has cis/trans isomerism, and so the compound must contain a C=C group with two different groups on each carbon atom.

(iii) In these compounds, you must have two hydrogen atoms on one of the carbon atoms.

(b) (i) Remember that nucleophiles must be defined in terms of a lone pair.

(iii) Do not use the word substitutes in your definition of substitution.

(a) (i)

	%	$\div A_r$		$\div 1.31$	empirical formula
carbon	47.1	÷12	= 3.93	= 3	C_3H_5Cl
hydrogen	6.5	÷1	= 6.5	= 5	
chlorine	46.4	÷35.5	= 1.31	= 1	

(ii) 

(iii) 

(b) **(i)** A nucleophile is a substance which uses its lone pair of electrons to form a covalent bond. ✓

It attacks electron deficient centres. ✓

(ii) Any one of OH^-, NH_3 or CN^-. ✓

(iii) When one atom or group replaces another in a molecule. ✓

Q3 How to score full marks

The mechanism for the free radical substitution reaction of chlorine and methane is: ✓

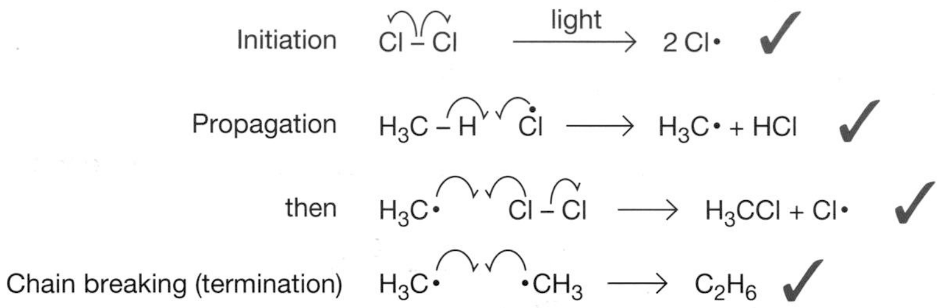

The mechanism for the electrophilic addition reaction of chlorine with ethene is: ✓

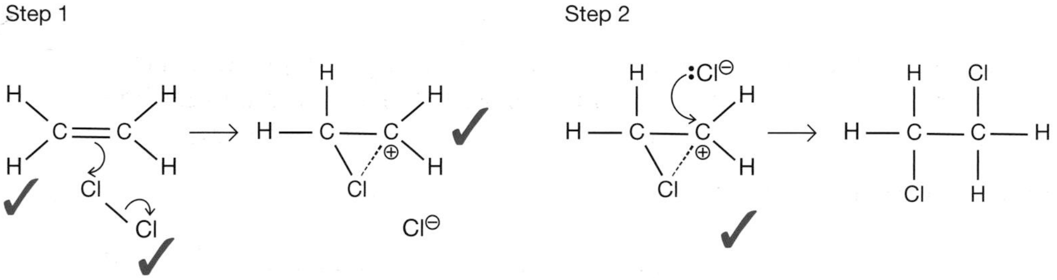

The Cl–Cl bond breaking in the free radical reaction with methane is an example of homolytic fission. ✓
In the addition reaction the bond breaking is heterolytic fission. ✓

> **Examiner's comments**
>
> A fish hook arrow  represents the movement of a single electron. The initiation step produces the radicals that start the reaction. In the propagation steps **one radical always produces one radical**. The termination step involves **two radicals forming a molecule.**
>
> A curly arrow  represents the movement of a **pair** of electrons. The arrow must start on a bond or atom and go to an atom to form a bond or an ion.
>
> The fission of the Cl–Cl bond in the reaction with methane results in **one** electron going to **each** chlorine atom. The heterolytic fission in the ethene reaction involves the σ bond pair **both** going to **one** of the two chlorine atoms.

Practice AS Level Exam Paper

Time allowed: $1\frac{1}{2}$ hours

1 (a) State the type of bonding in:

(i) sodium chloride **(ii)** silicon tetrafluoride

type .... type .... [2 marks]

(ii) Explain why aluminium fluoride is ionic and boils at 1270 °C whereas anhydrous aluminium bromide is covalent and boils at 265 °C.

....

.... [6 marks]

(b) Silicon is an element in the *p*-block of the Periodic Table.

(i) Using 1s, 2s, 2p … notation, write the electronic structure of silicon.

.... [2 marks]

(ii) State the number of unpaired electrons in one atom of silicon.

.... [1 mark]

(iii) Silicon has three naturally occurring isotopes, ^{28}Si, ^{29}Si and ^{30}Si. Explain why its relative atomic mass is 28.1 not 29.0.

.... [2 marks]

2 (a) 25.0 cm^3 of a 0.111 mol dm^{-3} solution of sodium carbonate, Na_2CO_3, was pipetted into a conical flask and two drops of phenol phthalein indicator added. This solution required 21.3 cm^3 of a 0.130 mol dm^{-3} solution of hydrochloric acid, HCl, for neutralization.

(i) Calculate the moles of sodium carbonate taken.

.... [2 marks]

(ii) Calculate the moles of hydrochloric acid needed.

.... [2 marks]

(iii) Hence, calculate the mole ratio of sodium carbonate to hydrochloric acid.

.... [1 mark]

(iv) Based on these results, suggest the equation for the reaction.

.... [2 marks]

(b) Hydrogen chloride gas, HCl(g), does not react with concentrated sulphuric acid, but when hydrogen bromide gas is bubbled into concentrated sulphuric acid, a brown gas is produced.

(i) Write the equation for the reaction between hydrogen bromide and sulphuric acid.

.... [2 marks]

(ii) Explain this reaction in terms of redox.

.... [4 marks]

(iii) Why is there no reaction between hydrogen chloride and sulphuric acid?

.... [1 mark]

3 2-bromopropane can be hydrolysed by warming with aqueous sodium hydroxide solution. They react according to the equation:

$$CH_3CHBrCH_3 + NaOH \rightarrow CH_3CH(OH)CH_3 + NaBr$$

2-bromopropane has a boiling point of 59 °C and is slightly soluble in water. It is not flammable. Propan-2-ol has a boiling point of 82 °C and is very soluble in water. It is very flammable.

(a) When 4.92 g of 2-bromopropane was warmed at 40 °C with an excess of a 1.0 mol dm^{-3} aqueous solution of sodium hydroxide for 20 minutes, 1.80 g of propan-2-ol was produced.

(i) Calculate the percentage yield.

....

.... [4 marks]

(ii) State two ways by which the rate of this reaction could be increased.

.... [2 marks]

(iii) Give two reasons why you would use a flask fitted with a reflux condenser rather than an open beaker for this reaction.

....

.... [2 marks]

(b) Explain why propan-2-ol:

(i) Has a higher boiling point than 2-bromopropane.

.... [4 marks]

(ii) Is more soluble in water than 2-bromopropane.

.... [2 marks]

4 (a) (i) Define standard enthalpy of combustion.

.... [3 marks]

(ii) Illustrate your answer with an equation that represents the standard enthalpy of combustion of ethane, $C_2H_6(g)$.

.... [2 marks]

(b) The standard enthalpies of combustion of ethene, $C_2H_4(g)$, ethane, $C_2H_6(g)$ and hydrogen, $H_2(g)$ are – 1409 kJ mol^{-1}, – 1560 kJ mol^{-1} and – 286 kJ mol^{-1}. Draw a Hess's Law diagram and use it to show that for the reaction:

$$C_2H_4(g) + H_2(g) \rightarrow C_2H_6(g)$$

the value of $\Delta H^{\ominus}$ equals – 135 kJ mol^{-1}.

[4 marks]

(c) Explain what factors determine the bond angles in a covalent compound.

.... [3 marks]

(d) Hence predict the CCH bond angle in:

(i) ethane

.... [3 marks]

(ii) ethene.

.... [3 marks]

5 (a) Both propane and propene react with bromine, but the conditions and the type of reaction are quite different, as propane reacts by free radical substitution and propene by electrophilic addition. Explain the meaning of the terms:

(i) free radical substitution

.... [2 marks]

(ii) electrophilic addition.

.... [4 marks]

(b) Propene is manufactured in large quantities by the cracking of a fraction from the distillation of crude oil. Its main use is in the manufacture of poly(propene). Draw the structure of a fragment of poly(propene) showing two repeat units.

.... [2 marks]

(c) The addition of bromine to propene produces $CH_3CHBrCH_2Br$. Draw the three other isomers of this product.

[3 marks]

(d) A sample of a halogenoalkane was heated under reflux with aqueous sodium hydroxide for five minutes. Describe tests that you would do to the resulting solution to show that the original halogenoalkane was a bromo-compound.

....

.... [4 marks]

6 (a) (i) Write the equation (with state symbols) which represents the first ionization energy of nitrogen.

.... [2 marks]

(ii) Explain why there is a big jump in values between the fifth and sixth ionization energies of nitrogen.

.... [3 marks]

(b) Nitrogen dioxide, NO_2, is a dark brown gas and is in equilibrium with dinitrogen tetroxide, N_2O_4, a pale almost colourless gas.

$$2NO_2(g) \rightleftharpoons N_2O_4(g)$$

(i) The colour changes from pale to dark brown as the mixture is heated. Deduce whether the reaction as written is exothermic or endothermic.

.... [3 marks]

(ii) Explain what would happen to the colour of the mixture when the pressure is increased at constant temperature.

.... [3 marks]

AS Sample Paper answers

Q1 How to score full marks

(a) **(i)** sodium chloride is ionic. ✓

silicon tetrafluoride is covalent. ✓

(ii) aluminium fluoride is ionic because the fluoride ion is so small✓ that it is not polarized✓ by the Al^{3+} ion. It boils at a high temperature because the force between the ions is very strong. ✓

Aluminium bromide is covalent because the much larger✓ bromide ion is polarized by the aluminium ion which has so a high charge density✓ that the bond becomes covalent. It has a much lower boiling temperature because the van der Waals forces✓ between the molecules are not very strong.

(b) **(i)** **Either** $1s^2, 2s^2\ 2p^6$✓ $3s^2\ 3p^2$✓

or [Ne]✓ $3s^2\ 3p^2$✓

(ii) It has two✓ unpaired electrons.

(iii) The % of the ^{29}Si and ^{30}Si isotopes is very small✓, therefore the average✓ mass comes out to 28.1.

Total 13 marks

Q2 How to score full marks

(a) **(i)** the amount of sodium carbonate

$= 0.111 \times \frac{25.0}{1000}$✓ $= 0.002775$ mol ✓

(ii) the amount of hydrochloric acid

$= 0.130 \times \frac{21.3}{1000}$✓ $= 0.002769$ mol ✓

(iii) the ratio is 1:1 ✓

(iv) $Na_2CO_3 + HCl \rightarrow NaHCO_3$✓ $+ NaCl$✓

(b) **(i)** $2HBr + H_2SO_4 \rightarrow Br_2$✓ $+ SO_2 + 2H_2O$
balanced equation✓

(ii) The oxidation number of bromine has gone up from –1 to 0✓ and so it has been oxidized✓. The oxidation number of sulphur has gone down from +6 to +4✓ and so it has been reduced.✓

(iii) HCl is not a strong enough reducing agent✓ to be reduced by the concentrated sulphuric acid.

Total 14 marks

Q3 How to score full marks

(a) **(i)** amount of 2-bromopropane

$= \frac{4.92\text{ g}}{123.0\text{ g mol}^{-1}} = 0.0400$ mol✓

theoretical amount of propan-2-ol
$= 0.0400$ mol✓

theoretical mass of propan-2-ol
$= 0.0400 \text{ mol} \times 60 \text{ g mol}^{-1} = 2.40$ g✓

% yield $= 100 \times \frac{1.80}{2.40} = 75.0$ %✓

(ii) Increase the temperature✓ (but not above 59 °C).

Increase the concentration of the sodium hydroxide solution.✓

(iii) To prevent the 2-bromopropane (or the organic compounds) from boiling off.✓

To prevent the propan-2-ol from catching fire.✓

(b) **(i)** In propan-2-ol there is hydrogen bonding✓ between the δ+ hydrogen of the OH group in one molecule and the δ- oxygen in another molecule✓, but the intermolecular forces in 2-bromopropane are only the weaker✓ dipole/dipole and induced dipole/dipole (dispersion) forces.✓

(ii) Propan-2-ol can form hydrogen bonds with the water✓ whereas 2-bromopropane cannot.✓

Total 14 marks

Q4 How to score full marks

(a) **(i)** It is the enthalpy change when 1 mol✓ of a substance is completely✓ burnt in oxygen under standard conditions (1 atm pressure and a stated temperature, usually 298K)✓

(ii) $C_2H_6(g) + 3\frac{1}{2}\ O_2(g) \rightarrow 2CO_2(g) + 3H_2O(l)$

Balanced equation for 1 mol✓ state symbols✓ (note that water is liquid)

(b)

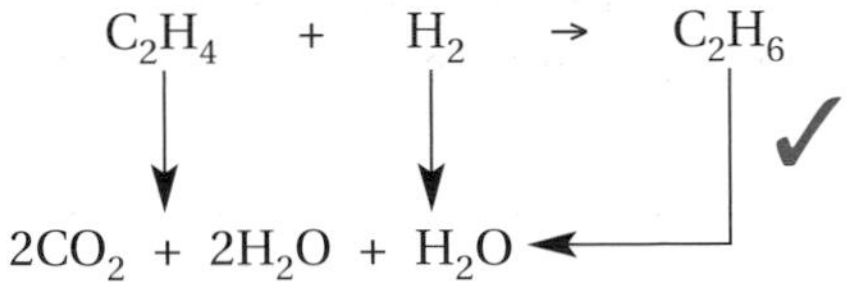

$\Delta H_r = \Delta H_c$ (ethene) $+ \Delta H_c$ (hydrogen) –
ΔH_c (ethane)✓ (or from labelled Hess's law diagram)

$= (-1409) + (-286) - (-1560)$✓ $= -135$ kJ mol^{-1}✓

(c) It is determined by the number of pairs of electrons around the atom✓ which repel to get as far apart as possible✓, but lone pair/lone pair repulsion is greater than lone pair/bond pair which in turn is greater than bond pair/bond pair.✓

(d) (i) Ethane: 4 separate bond pairs✓ arrangement is tetrahedral✓ angle $109\frac{1}{2}$°.✓

(ii) Ethene: 3 separate bond pairs✓ arrangement square planar✓ angle 120°.✓

Total 18 marks

Q5 How to score full marks

(a) (i) A free radical is an atom with an **unpaired** electron.✓

Substitution is when one species (atom, ion or group) replaces another.✓

(ii) An electrophile is a species which will accept a pair of electrons✓ to form a covalent bond✓. It attacks electron rich (δ-) sites.✓

Addition is when two substances form a single compound.✓

(b) —$CH(CH_3)$—CH_2—$CH(CH_3)$—CH_2—
carbon skeleton✓ continuation bonds✓

(c) $CHBr_2CH_2CH_3$✓ $CH_2BrCH_2CH_2Br$✓ $CH_3CBr_2CH_3$✓

(d) **Either**

acidify with dilute nitric acid✓
add silver nitrate solution✓
bromo compounds give a cream (pale yellow) ppt✓ which is insoluble in dilute ammonia but soluble in concentrated ammonia✓

or

acidify with dilute nitric acid✓
add chlorine water✓
shake with hexane✓ (or another immiscible organic solvent)
bromo compounds give a red/brown colour to the hexane layer✓

Total 15 marks

Q6 How to score full marks

(a) (i) $N(g) \rightarrow N^+(g) + e^-$ equation✓ state symbols.✓

(ii) The sixth electron comes from an inner shell (or 1st orbit)✓ so is nearer to the nucleus✓ and is not shielded by inner electrons, so is held on more firmly.✓

(b) (i) The equilibrium shifts to the left (towards dark brown NO_2).✓

An increase in temperature shifts the equilibrium in the endothermic direction.✓

So the reaction, as written, is exothermic.✓.

(ii) The equilibrium moves to the side with the fewer gas molecules✓ that is to the right,✓ therefore the gas gets paler.✓

Total 11 marks

TOTAL MARK FOR THE PAPER = 85